D0153795

LATENT VARIABLE MODELS
An Introduction to Factor, Path, and Structural Analysis

John C. Loehlin
University of Texas at Austin

LATENT VARIABLE MODELS
An Introduction to Factor, Path, and Structural Analysis

LEA LAWRENCE ERLBAUM ASSOCIATES, PUBLISHERS

1987   Hillsdale, New Jersey   London

Copyright © 1987 by Lawrence Erlbaum Associates, Inc.
  All rights reserved. No part of this book may be reproduced in
  any form, by photostat, microform, retrieval system, or any other
  means, without the prior written permission of the publisher.

Lawrence Erlbaum Associates, Inc., Publishers
365 Broadway
Hillsdale, New Jersey 07642

Library of Congress Cataloging-in-Publication Data

Loehlin, John C.
  Latent variable models.

  Bibliography: p.
  Includes index.
  1. Latent variables.   2. Latent structure analysis.
3. Factor analysis.   4. Path analysis.   I. Title.
QA278A.6.L64   1987   105'.72   86-29294
ISBN 0-89859-963-6
ISBN 0-89859-965-2 (pbk.)

Printed in the United States of America
10 9 8 7 6 5 4 3 2

Contents

Contents

Contents

Contents

Preface

This book is intended as an introduction to an exciting growth area in social science methodology--the use of multiple-latent-variable models. Psychologists and other social scientists have long been familiar with one subvariety of such modeling, factor analysis--more properly, exploratory factor analysis. In recent decades, confirmatory factor analysis, path analysis, and structural equation modeling have come out of specialized niches and are making their bid to become basic tools in the research repertoire of the social scientist, particularly the one who is forced to deal with complex real-life phenomena in the round: the sociologist, the political scientist, the social, educational, clinical, industrial, personality or developmental psychologist, the marketing researcher, and the like.

All these methods are at heart one, as I have tried to emphasize in the chapters to follow. I have used preliminary versions of this book in teaching graduate students from psychology and related disciplines, and have found the particular approach used--via path diagrams--to be effective in helping not-too-mathematical students grasp underlying relationships, as opposed to merely going through the motions of running computer programs. In some sections of the book a certain amount of elementary matrix algebra is employed; an appendix on the topic is provided for those who may need help here.

In the interests of accessability, I have tried to maintain a relatively informal style, and to keep the main text fairly uncluttered with references. The notes at the end of each chapter are intended to provide the serious student with a path into the technical literature, as well as to draw his or her attention to some issues beyond the scope of the basic treatment.

The book is not closely tied to a particular computer program or package, although there is some special attention paid to LISREL. I assume that most users will have access to a latent-variable model-fitting program on the order of LISREL, COSAN, EQS, or MILS, and an exploratory factor analysis package such as those in SPSS, BMDP, or SAS. In some places, a matrix manipulation facility such as that in MINITAB or SAS would be helpful. I have provided some introductory material but have not tried to tell students all they need to know to run actual programs--such information is often local, ephemeral, or both. The instructor should expect to provide some handouts and perhaps a bit of hands-on assistance in getting students started. The reader going it on his or her own will require access to current manuals for the computer programs to be used.

Finally, it gives me great pleasure to acknowledge the help and encouragement that others have provided. Perhaps first credit should go to the students who endured earlier versions of the manuscript and cheerfully pointed out various errors and obscurities. These brave pioneers included Mike Bailey, Cheryl Beauvais, Alan Bergman, Beth Geer, Steve Gregorich, Priscilla Griffith, Jean Hart, Pam Henderson, Wes Hoover, Vivian Jenkins, Tock Lim, Scott Liu, Jacqueline Lovette, Frank Mulhern, Steve Predmore, Naftali Raz, and Lori Roggman. Among other colleagues who have been kind enough to read and comment on various parts of the manuscript are Carole Holahan, Phil Gough, Maria Pennock-Roman, Peter Bentler, and several anonymous reviewers. I am especially grateful to Jack McArdle for extensive comments on the manuscript as a whole, and to Jack Cohen for his persuasive voice with the publishers. Of course, these persons should not be blamed for any defects that may remain. For one thing, I didn't always take everybody's advice.

I am grateful to the University of Chicago, to *Multivariate Behavioral Research,* and to the Macmillan Publishing Company for permission to reprint or adapt published materials, and to the many previous researchers and writers cited in the book--or, for that matter, not cited--whose contributions have defined this rapidly developing and exciting field.

Finally, I owe a special debt to the members of my family: Jennifer and James, who worked their term papers in around my sessions at the Macintosh, and Marj, who provided unfailing support throughout.

J. C. L.

Chapter One:
Path Models in Factor, Path, and
Structural Analysis

Scientists dealing with behavior, especially those who observe it occurring in its natural settings, rarely have the luxury of the simple bivariate experiment, in which a single independent variable is manipulated and the consequences observed for a single dependent variable. Even those scientists who think they do are often mistaken: The variables they directly manipulate and observe typically are not the ones of real theoretical interest but are merely some convenient variables acting as proxies or indexes for them. A full experimental analysis would again turn out to be multivariate, with a number of alternative experimental manipulations on the one side, and a number of alternative response measures on the other.

Over many years numerous statistical techniques have been developed for dealing with situations in which multiple variables, some unobserved, are involved. Such techniques often involve heroic amounts of computation. Until the advent of powerful digital computers and associated software, the use of these methods tended to be restricted to the dedicated few. But in the last couple of decades it has been feasible for any interested behavioral scientist to take a multivariate approach to his or her data. Many have done so. The explosive growth in the use of computer software packages such as SPSS, BMDP, SAS, LISREL, and others is one evidence of this. All these packages contain programs capable of doing one form or another of multivariate analysis involving unobserved variables.

The common features of the methods discussed in this book are that (a) multiple variables--three or more--are involved, and that (b) one or more of these variables is unobserved, or *latent*. Neither of these criteria provides a decisive boundary. Bivariate methods may often be regarded as special cases of multivariate methods. Some of the methods we discuss can be--and often are--applied in situations where all the variables are in fact observed. Nevertheless, the main focus of our interest is on what we call, following Bentler (1980), *latent variable analysis*, a term encompassing such specific methods as factor analysis, path analysis, structural equation analysis, multidimensional scaling, and so on, all of which share these defining features.

Path Diagrams

An easy and convenient representation of the relationships among a number of variables is the *path diagram.* In such a diagram we use capital letters, A, B, X, Y, and so on, to represent variables. The connections among variables are represented in path diagrams by two kinds of arrows: a straight, one-headed arrow represents a causal relationship between two variables, and a curved two-headed arrow represents a simple correlation between them.

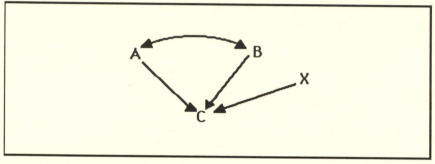

Fig. 1.1 Example of a simple path diagram.

Figure 1.1 shows an example of a path diagram. Variables A, B, and X each is assumed to have a causal effect on variable C. Variables A and B are assumed to be correlated with each other. Variable X is assumed to affect C but to be uncorrelated with either A or B. Variable C might (for example) represent young children's intelligence. Variables A and B could represent father's and mother's intelligence, assumed to have a causal influence on their child's intelligence. (The diagram is silent as to whether this influence is environmental, genetic, or both.) The curved arrow between A and B allows for the likely possibility that father's and mother's intelligence will be correlated. Arrow X represents the fact that there are other variables, independent of mother's and father's intelligence, that can affect a child's intelligence.

Figure 1.2 shows another example of a path diagram. T is assumed to affect both A and B, and each of the latter variables is also affected by an additional variable; these are labeled U and V, respectively. This path diagram could represent the reliability of a test, as described in classical psychometric test theory. A and B would stand (say) for scores on two alternate forms of a test. T would represent the unobserved true score on the trait being measured, which is assumed to affect the observed scores on both forms of the test. U and V would represent factors specific to each form of the test or to the occasions on

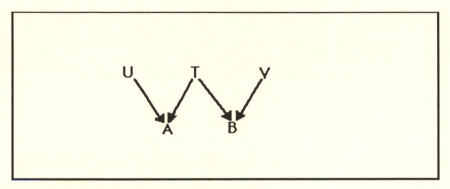

Fig. 1.2 Another path diagram: test reliability.

which it was administered, which would affect any given performance but be unrelated to the true trait. (In classical psychometric test theory, the variance in A and B resulting from the influence of T would be called *true score variance* and that caused by U or V would be called *error variance*. The proportion of the variance of A or B due to T would be called the *reliability* of the test.)

Figure 1.3 shows a path representation of events over time. In this case, the capital letters A and B are used to designate two variables, with subscripts to identify the occasions on which they are measured: Both A and B are measured at time 1, A is measured again at time 2, and B at time 3. In this case, the diagram indicates that both A_1 and B_1 are assumed to affect A_2, but that the effect of A_1 on B at time 3 is wholly via A_2--there is no direct arrow drawn leading from A_1 to B_3. It is assumed that A_1 and B_1 are correlated, and that A_2 and B_3 are subject to additional influences independent of A and B, here

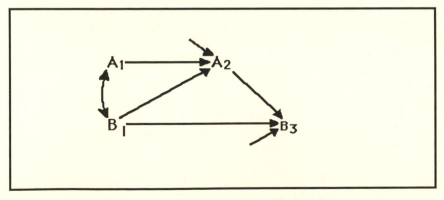

Fig. 1.3 A path diagram involving events over time.

represented by short, unlabeled arrows.. These additional influences could have been labeled, say, X and Y, but are often left unlabeled in path diagrams, as here, to indicate that they refer to other, unspecified influences on the variable to which they point. Such arrows are called *residual arrows* to indicate that they represent causes residual to those explicitly identified in the diagram.

The meaning of "cause" in a path diagram

Straight arrows in path diagrams are said to represent *causal relationships*--but in what sense of the sometimes slippery word "cause" ? In fact, we do not need to adopt any strict or narrow definition of cause in this book, because path diagrams can be--and are--used to represent causes of various different kinds, as the examples we have considered suggest. The essential feature for the use of a causal arrow in a path diagram is the assumption that a change in the variable at the tail of the arrow will result in a change in the variable at the head of the arrow, all else being equal (i.e., with all other variables in the diagram held constant). Note the one-way nature of this process--imposing a change on the variable at the head of the arrow does *not* bring about a change in the tail variable. A variety of common uses of the word "cause" can be expressed in these terms, and hence can legitimately be represented by a causal arrow in a path diagram.

Completeness of a path diagram

Variables in a path diagram may be grouped in two classes: those that do not receive causal inputs from any other variable in the path diagram, and those that receive one or more such causal inputs. Variables in the first of these two classes are referred to as *exogenous, independent,* or *source variables.* Variables in the second class are called *endogenous, dependent,* or *downstream variables. Exogenous variables* (Greek: "of external origin") are so called because their causal sources lie external to the path diagram; they are causally *independent* with respect to other variables in the diagram-- straight arrows may lead away from them but never toward them. These variables represent causal *sources* in the diagram. Examples of such source variables in Fig. 1.3 are A_1, B_1, and the two unlabeled residual variables.

Endogenous variables ("of internal origin") have causal sources that lie within the path diagram; they are causally *dependent* on other variables--one or more straight arrows lead into them. These variables lie causally *downstream* from source variables. Examples of downstream variables in Fig. 1.3 are A_2 and B_3.

In Fig. 1.2, U, T, and V are source variables, and A and B are downstream variables. Look back at Fig. 1.1. Which are the source and downstream variables in this path diagram? (I hope you identified A, B, and X as source variables, and C as downstream.)

In a proper and complete path diagram, all the source variables are interconnected by curved arrows, to indicate that they may be intercorrelated-- unless it is explicitly assumed that their correlation is zero, in which case the curved arrow is omitted. Thus the absence of a curved arrow between two source variables in a path diagram, as between X and A in Fig. 1.1, or T and U in Fig. 1.2, is not an expression of ignorance but an explicit statement about assumptions underlying the diagram.

Downstream variables, on the other hand, are never connected by curved arrows in path diagrams. (Actually, some authors use downstream curved arrows as a shorthand to indicate correlations among downstream variables caused by other variables than those included in the diagram: We use correlations between residual arrows for this purpose, which is consistent with our convention because the latter are source variables.) Residual arrows point at downstream variables, never at source variables. Completeness of a path diagram requires that a residual arrow be attached to every downstream variable unless it is explicitly assumed that all the causes of variation of that variable are included among the variables upstream from it in the diagram. (This convention is also not universally adhered to: Occasionally, path diagrams are published with the notation "residual arrows omitted." This is an unfortunate practice because it leads to ambiguity in interpreting the diagram: Does the author intend that all the variation in a downstream variable is accounted for within the diagram, or not?)

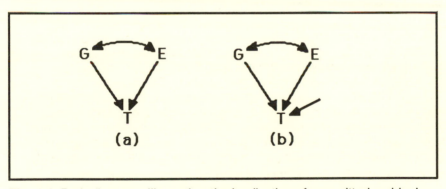

Fig. 1.4 Path diagrams illustrating the implication of an omitted residual arrow.

Figure 1.4 shows an example in which the presence or absence of a residual arrow makes a difference. The source variables G and E refer to the genetic and environmental influences on a trait T. The downstream variable T in Fig. 1.4 (a) has no residual arrow. That represents the assumption that the variation of T is completely explained by the genetic and environmental influences upon it. This is a theoretical assumption that one might sometimes

5

wish to make. Fig. 1.4 (b), however, represents the assumption that genetic and environmental influences are not sufficient to explain the variation of T--some additional factor or factors, perhaps measurement error or gene-environment interaction-- may need to be taken into account in explaining T. Obviously, the assumptions in Figures 1.4 (a) and (b) are quite different, and one would not want it assumed that (a) was the case when in fact (b) was intended.

Finally, all significant direct causal connections between source and downstream variables, or between one downstream variable and another, should be included as straight arrows in the diagram. Omission of an arrow between A_1 and B_3 in Fig. 1.3 is a positive statement: that A_1 is assumed to affect B_3 only by way of A_2.

The notion of completeness in path diagrams should not be taken to mean that the ideal path diagram is one containing as many variables as possible connected by as many arrows as possible. Exactly the opposite is true. The smallest number of variables connected by the smallest number of arrows that can do the job is the path diagram to be sought for, because it represents the most parsimonious explanation of the phenomenon under consideration. Big, messy path diagrams are likely to give trouble in many ways. Nevertheless, often the simplest explanation of an interesting behavioral or biological phenomenon does involve causal relationships among a number of variables, not all observable. A path diagram provides a way of representing in a clear and straightforward fashion what is assumed to be going on in such a case.

Notice that most path diagrams could in principle be extended indefinitely back past their source variables: These could be taken as downstream variables in an extended path diagram, and the correlations among them explained by the linkages among their own causes. Thus, the parents in Fig. 1.1 could be taken as children in their own families, and the correlation between them explained by a model of the psychological and sociological mechanisms that result in mates having similar IQs. Or in Fig. 1.3, one could have measured A and B at a preceding time zero, resulting in a diagram in which the correlation between A_1 and B_1 is replaced by a superstructure of causal arrows from A_0 and B_0, themselves probably correlated. There is no hard and fast rule in such cases, other than the general maxim that simpler is better, which usually means that if going back entails multiplying variables, do not do it unless you have to. Sometimes, of course, you have to, when some key variable lies back upstream.

Other assumptions in path diagrams

It is assumed in path diagrams that causes are unitary, that is, in a case such as Fig. 1.2, that it is meaningful to think of a single variable T that is the cause of A and B, and not (say) two separate and distinct aspects of a phenomenon T, one of which causes A and one B. In the latter case, a better representation would be to replace T by two different (possibly correlated) variables.

An exception to the rule of unitary causes is residual variables, which typically represent multiple causes of a variable that are external to the path diagram. Perhaps for this reason, path analysts do not always solve for the path coefficients associated with the residual arrows in their diagrams. It is, however, good practice to solve at least for the proportion of variance associated with such residual causes (more on this later). It is nearly always useful to know what proportion of the variation of each downstream variable is accounted for by the causes explicitly included within the path diagram, and what proportion is not.

Another assumption made in path diagrams is that the causal relationships represented by straight arrows are linear. This is usually not terribly restricting--mild departures from linearity are often reasonably approximated by linear relationships, and if not, it may be possible to transform variables so as to linearize their relationships with other variables. The use of log income, rather than income, or reciprocals of latency measures, or arcsine transformations of proportions, would be examples of transformations often used by behavioral scientists for this purpose. In drawing a path diagram, one ordinarily does not have to worry about such details--one can always make the blanket assumption that one's variables are measured on scales for which relationships are reasonably linear. But in evaluating the strength of causal effects with real data, the issue of nonlinearity may arise. If variable A has a positive effect on variable B in part of its range and a negative effect in another, it is hard to assign a single number to represent the effect of A on B. However, if A is suitably redefined, perhaps as an absolute deviation from some optimum value, this may be possible. In Chapter 7 we consider some direct approaches to dealing with nonlinear relationships of latent variables.

A note on the orientation of path diagrams

Path diagrams are ordinarily arranged on the page with the flow of causation going more or less from left to right or from top to bottom, although occasionally one sees other arrangements, such as a path diagram radiating out from the center. Figures 1.1, 1.2, and 1.4 are examples of top-to-bottom orientation, Fig. 1.3 of left-to-right. Any of these is perfectly acceptable, though sociologists seem to have some preference for the left-to-right arrangement and geneticists for top-to-bottom. We use both in this book, with the choice often determined by such trivial considerations as how well the diagram fits on the page.

Feedbacks and mutual influences

In our examples so far we have restricted ourselves to path diagrams in which, after the source variables, there was a simple downstream flow of causation-- no paths that loop back on themselves or the like. Most of the cases we consider in this book have this one-way causal flow, but path representations can be used to deal with more complex situations involving causal loops, as we see later in Chapter 4. Examples of two such non-one-way cases are shown in Fig. 1.5. In Fig. 1.5 (a) there is a mutual causal influence between variables C and D: each affects the other. A causal sequence could go from A to C to D to C to D again and so on. In Fig. 1.5 (b) there is an extended feedback loop: A affects B which affects C which in turn affects A.

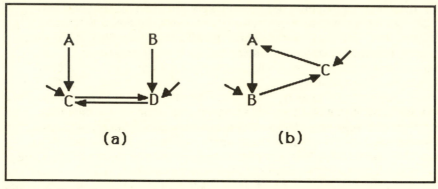

Fig. 1.5 Path diagrams with (a) mutual influences and (b) a feedback loop.

Direct and indirect causal paths

Sometimes it is useful to distinguish between *direct* and *indirect* causal effects in path diagrams. A direct effect is represented by a single causal arrow between the two variables concerned. In Fig. 1.5 (b) variable B has a direct effect on variable C. There is a causal arrow leading from B to C. If B is changed we expect to observe a change in C. Variable A, however, has only an indirect effect on C because there is no direct arrow from A to C. There is, however, an indirect causal effect transmitted via variable B. If A changes, B will change, and B's change will affect C, other things being equal. Thus, A can be said to have a causal effect on C, although an indirect one. In Fig. 1.5 (a) variable B has a direct effect on variable D, an indirect effect on variable C, and no causal effect at all on variable A.

8

Path Analysis

Path diagrams are useful enough as simple descriptive devices, but they can be much more than that. Starting from empirical data, one can solve for a numerical value of each curved and straight arrow in a diagram to indicate the relative strength of that correlation or causal influence. Numerical values, of course, imply scales on which they are measured. For most of this chapter we assume that all variables in the path diagram are expressed in standard score form, that is, with a mean of zero and a standard deviation of one. Covariances and correlations are thus identical. This simplifies matters of presentation, and is a useful way of proceeding in many practical situations. Later, we see how the procedures can be applied to data in original raw score units, and consider some of the situations in which this approach is to be preferred. We also assume for the present that we are dealing with unlooped path diagrams.

The steps of constructing and solving path diagrams are referred to collectively as *path analysis,* a method originally developed by the American geneticist Sewall Wright as early as 1920, but only extensively applied in the social and behavioral sciences during the last couple of decades.

Wright's rules

Briefly, Wright showed that if a situation can be presented as a proper path diagram, then the correlation between any two variables in the diagram can be expressed as the *sum of the compound paths connecting these two points,* where a compound path is a path along arrows that follows three rules:

      (a) no loops;
      (b) no going forward then backward;
      (c) a maximum of one curved arrow per path.

The first rule means that a compound path must not go twice through the same variable. In Fig. 1.6 (a) the compound path ACF would be a legitimate path between A and F, but the path ACDECF would not be because it involves going twice through variable C.

The second rule means that on a particular path, after one has once gone forward along one or more arrows, it is not legitimate to proceed backwards along others. (Going backward first and then forward is, however, quite proper.) In Fig. 1.6 (b) the compound path BAC is a legitimate way to go from B to C, the path BDC is not. In the former, one goes backward along an arrow (B to A) and then forward (A to C), which is allowable, but path BDC would require going forward then backward, which is not. This asymmetry may seem a bit less arbitrary if one realizes that it serves to permit events in the diagram to be connected by common causes (A), but not by common consequences (D).

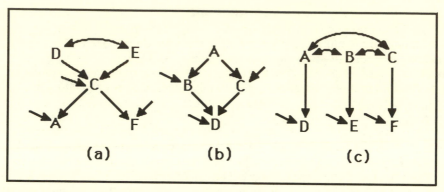

Fig. 1.6 Illustrations of Wright's rules.

The third rule is illustrated in Fig. 1.6 (c). DACF is a legitimate compound path between D and F; DABCF is not, because it would require traversing two curved arrows. Likewise, DABE is a legitimate path between D and E, but DACBE is not.

   Figure 1.7 serves to provide examples of tracing paths in a path diagram according to Wright's rules. This figure incorporates three source variables, A, B, and C, and three downstream variables, D, E, and F. We have designated each arrow by a lower case letter for convenience in representing compound paths. Each lower case letter stands for the *value* or magnitude of the particular causal effect or correlation. A simple rule indicates how these values are combined: *The numerical value of a compound path is equal to the product of the values of its constituent arrows.* So that simply writing the lower case letters of a path in sequence is at the same time writing an expression for the numerical value of that path.

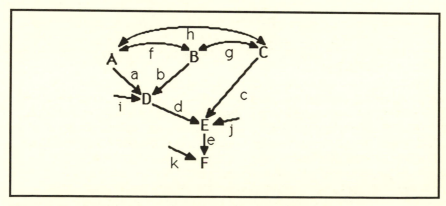

Fig. 1.7 Examples of tracing paths in a path diagram.

For example, what is the correlation between variables A and D in Fig. 1.7? Two paths are legal: *a* and *fb*. A path like *hgb* would be excluded by the rule about only one curved arrow, and paths going further down the diagram like *adcgb* would violate both the rules about no forward then backward and no loops. So the numerical value of r_{AD} can be expressed as *a* +*fb*. I hope that the reader can see that $r_{BD} = b$ +*fa,* and that $r_{CD} = gb$ +*ha.*

What about r_{AB}? Just *f.* Path *hg* would violate the third rule, and paths like *ab* or *adcg* would violate the second. It is, of course, quite reasonable that r_{AB} should equal *f,* because that is just what the curved arrow between A and B means. Likewise $r_{BC} = g$ and $r_{AC} = h$.

Let us consider a slightly more complicated case: r_{AE}. There are three paths: *ad, fbd,* and *hc.* Note that although variable D is passed through twice, this is perfectly legal, because it is only passed through once on any given path. You might wish to pause at this point to work out r_{BE} and r_{CE} for yourself. (I hope you got *bd* +*fad* +*gc* and *c* +*gbd* +*had*.)

Finally, consider the correlation between A and F. This works out to *ade* +*fbde* +*hce.* Notice that this can be rewritten as (*ad* +*fbd* +*hc*)*e*, or simply (r_{AE})*e* ; that is, once we have solved for the correlation r_{AE}, we can think of

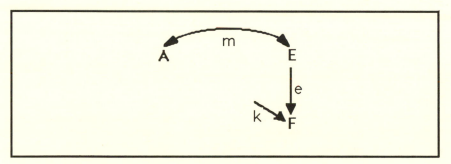

Fig. 1.8 A simplified version of Fig. 1.7.

the path diagram as having the simple form shown in Fig. 1.8, of two variables A and E whose correlation (call it now *m*) is known, plus a causal path *e* from variable E to F. In working with complex path diagrams, this technique of substituting a single value for a discrete subunit of the diagram is often helpful. Note that this simplification can be made because the causation of F is separable in the diagram from the network of paths connecting A and E. If there had also been a direct path from, say, B to F, this could not properly be represented in Fig. 1.8, and this particular simplification would not be possible.

You should now try your hand at the five remaining correlations in Fig.1.7:

r_{DE}, r_{EF}, r_{BF}, r_{CF}, and r_{DF}. The answers are not given until later in the chapter, to minimize the temptation of peeking at them first.

Numerical solution of a path diagram

Given that we can express each of the correlations among a set of observed variables in a path diagram as a sum of compound paths, can we reverse this process and solve for the values of the causal paths given the correlations? The answer is that often we can.

    Consider the example of Fig. 1.1, redrawn as Fig. 1.9. Recall that variables A and B were fathers' and mothers' intelligence, and C was children's intelligence. X is a residual variable, representing other unmeasured influences on child's intelligence that are independent of the parents'

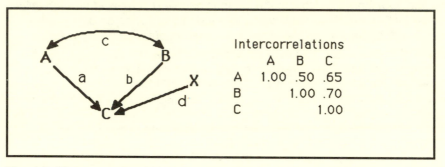

Fig. 1.9 The example of Fig. 1.1, with observed inter-correlations of A, B, and C.

intelligence. Suppose that in some suitable population of families we were to observe the correlations shown on the right in Fig. 1.9. We can now, using our newfound knowledge of path analysis (and ignoring X for the moment), write the following three equations:

$$r_{AB} = c$$
$$r_{AC} = a + cb$$
$$r_{BC} = b + ca.$$

Because we know the observed values r_{AB}, r_{AC}, and r_{BC}, we have three simultaneous equations in three unknowns:

$$c = .50$$
$$a + cb = .65$$
$$b + ca = .70.$$

The solution is straightforward. Substitution for c in the second and third equations yields two equations in two unknowns:

$$a + .50b = .65$$
$$.50a + b = .70.$$

Doubling the second of these equations and subtracting the first from it gives $1.50b = .75$, from whence $b = .50$. Substituting for b in either equation gives $a = .40$. Thus, if we were to observe the set of intercorrelations given in Fig. 1.9, *and if our causal model is correct,* we could conclude that the causal influences of fathers' and mothers' intelligence on child's intelligence could be represented by values of .40 and .50, respectively, for the causal paths a and b.

What do these numbers mean? They are, in fact, standardized partial regression coefficients--we call them *path coefficients* for short. Because they are *regression coefficients,* they tell us to what extent a change on the variable at the tail of the arrow is transmitted to the variable at the head of the arrow. Because they are *partial* regression coefficients, this is the change that occurs with all other variables in the diagram held constant. Because they are *standardized* partial regression coefficients, we are talking about changes measured in standard deviation units. Specifically, the value of .40 for a means that if we were to select fathers who were one standard deviation above the mean for intelligence--but keeping mothers at the mean--their offspring would average four-tenths of a standard deviation above the population mean. (Unless otherwise specified, we are assuming in this chapter that the numbers we deal with are population values, so that issues of statistical inference do not complicate the picture.)

Because paths a and b are standardized partial regression coefficients, also known in multiple regression problems as *beta weights,* one might wonder if we can solve for them as such, by treating the path analysis as a sort of multiple regression problem. The answer is, yes we can, at least in cases where all variables are measured. In the present example, A, B, and C are assumed known, so we can solve for a and b by considering this as a multiple regression problem in predicting C from A and B.

Using standard formulas (e.g., McNemar, 1969, p. 192):

$$\beta_1 = (.65 - .70 \times .50)/(1 - .50^2) = .40$$
$$\beta_2 = (.70 - .65 \times .50)/(1 - .50^2) = .50,$$

or exactly the same results as before.

Viewing the problem in this way, we can also interpret the squared multiple correlation between C and A and B as the proportion of the variance of

C which is accounted for by A and B jointly. In this case $R^2_{C \cdot AB} = \beta_1 r_{AC} + \beta_2 r_{BC} = .40 \times .65 + .50 \times .70 = .61$. Another way in which we can arrive at the same figure from the path diagram is by following a path-tracing procedure: The predicted variance of C is *the sum of the paths from* C *to itself via* A *or* B *or both which follow Wright's rules.* In this case there is the path to A and back, with value a^2, the path to B and back, with value b^2, and the two paths *acb* and *bca* : $.40^2 + .50^2 + 2 \times .40 \times .50 \times .50 = .16 + .25 + .20 = .61$.

Either of these approaches will also allow us to solve for the value of the path *d* which leads from the unmeasured residual X. The variance that A and B jointly account for is R^2, or .61. The variance that X accounts for is thus $1 - R^2$, that is, $1 - .61$, or .39. This, therefore, must be the square of the correlation between X and C. So the value of *d* is $\sqrt{.39}$, or .62.

So long as all variables are measured one can proceed to solve for the causal paths in a path diagram as beta weights in a series of multiple regression analyses. Thus, in Fig. 1.7 one could solve for *a* and *b* from the correlations among A, B, and D; for *d* and *c* from the correlations among D, C, and E; and for *e* as the simple correlation between E and F. The residuals *i*, *j*, and *k* can then be obtained as $\sqrt{(1 - R^2)}$ in the various multiple regressions.

In general, however, we must deal with path diagrams involving unmeasured, latent variables. We cannot directly calculate the correlations of these with observed variables, so a simple multiple regression approach does not work. We need, instead, to carry out some variant of the first approach--that is, to solve a set of simultaneous equations with at least as many equations as there are unknown values to be obtained.

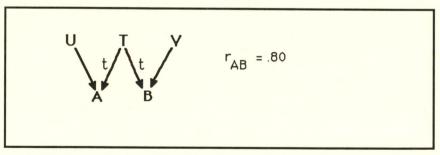

Fig. 1.10 The example of Fig. 1.2, with observed correlation of .80 between alternate forms A and B of a test.

Consider the example of Fig. 1.2, test reliability, repeated for convenience as Fig. 1.10. We wish to solve for the values of the causal paths between the true score T and the observed scores A and B. But T is an unobserved, latent variable; all we have is the observed correlation .80 between forms A and B of the test. How can we proceed? If we are willing to assume that A and B have

14

the same relation to T, which they should have if they are really parallel alternate forms of a test, we can write from the path diagram the equation

$$r_{AB} = t^2 = .80,$$

from which it follows that $t = \sqrt{.80} = .89$. It further follows that t^2 or 80% of the variance of each of the alternate test forms is attributable to the true score on the trait, that 20% is therefore due to error, and that the values of the residual paths from U and V are $\sqrt{.20}$ or .45.

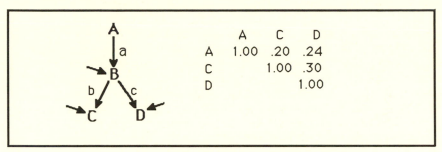

	A	C	D
A	1.00	.20	.24
C		1.00	.30
D			1.00

Fig. 1.11 Another simple path diagram with a latent variable.

Figure 1.11 presents another case of a path diagram containing a latent variable. It is assumed that A, C, and D are measured. Their intercorrelations are given to the right of the figure.   B is not measured, so we do not know its correlations with A, C, and D. We can, however, write 3 equations for the three known correlations in terms of the three paths *a, b,* and *c,* and (as it turns out) these three equations can be solved for the values of the three causal paths.
The equations are:

$$r_{AC} = ab$$
$$r_{AD} = ac$$
$$r_{CD} = bc.$$

A solution is:

$$r_{AC}\, r_{CD}/r_{AD} = ab \times bc\,/ac\ = b^2 = .20 \times .30/.24 = .25; \ \ b = .50$$
$$a = r_{AC}/b\ = .20/.50 = .40$$
$$c = r_{AD}/a = .24/.40 = .60.$$

Note that another possible solution would be numerically the same, but with all paths negative, because b^2 also has a negative square root.

15

(By the way, to keep the reader in suspense no longer about the problem regarding Fig. 1.7: $r_{DE} = d + ahc + bgc$, $r_{EF} = e$, $r_{BF} = bde + fade + gce$, $r_{CF} = ce + gbde + hade$, $r_{DF} = de + ahce + bgce$.)

Under, over, and just-determined path diagrams

Figure 1.12 (a) shows another simple path diagram. It is somewhat like Fig. 1.11 upside down: Instead of one cause of the latent variable and two effects, there are now two causes and one effect.

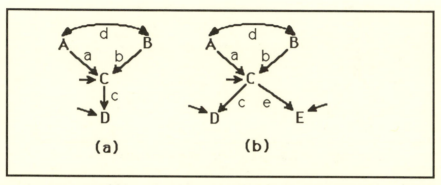

Fig. 1.12 Examples of under and overdetermined path diagrams.

However, this change has made a critical difference. There are still just three intercorrelations among the three observed variables A, B, and D, yielding three equations. But now there are four unknown values to be estimated: *a*, *b*, *c*, and *d*. One observed correlation, r_{AB}, estimates *d* directly. But that leaves only two equations, $r_{AD} = ac$ and $r_{BD} = bc$, to estimate the three unknowns *a*, *b*, and *c*, and no unique solution is possible. The path diagram is said to be *underdetermined*.

In the preceding problem of Fig. 1.11, there were three equations in three unknowns, and an exact solution was possible. Such a case is described as *just determined*. Fig. 1.12 (b) shows a third case, of an *overdetermined* path diagram. As in the left-hand figure, C is a latent variable and A and B are source variables, but an additional measured downstream variable E has been added. Now there are six observed intercorrelations among the observed variables A, B, D, and E, yielding six equations, whereas we have only added one unknown, giving five unknowns to be solved for. More equations than unknowns does not guarantee overdetermination, but in this case for most observed sets of correlations there will be no single solution for the unknowns that will satisfy all six equations simultaneously. What is ordinarily done in such cases is to seek values for the unknowns that come as close as possible to accounting for the observed intercorrelations (we defer until the next chapter a consideration of what "as close as possible" means).

It might be thought that just-determined path diagrams, because they permit exact solutions, would be the ideal to be sought for. But in fact for the behavioral scientist overdetermined path diagrams are usually to be preferred. The reason is that the data of the behavioral scientist typically contain both sampling and measurement error, and an exact fit to these data is an exact fit to the error as well as to the truth they contain. Whereas--if we assume that errors are more or less random--a best overall fit to the redundant data of an overdetermined path diagram will usually provide a better approximation to the underlying true population values. In addition, as we see later, overdetermined path diagrams may permit statistical tests of goodness of fit, which just-determined diagrams do not.

Factor Models

An important subdivision of latent variable analysis is traditionally known as factor analysis. In recent discussions of factor analysis, a distinction is often drawn between *exploratory* and *confirmatory* varieties. In exploratory factor analysis, which is what is usually thought of as "factor analysis" if no qualification is attached, one seeks under rather general assumptions for a latent variable structure that could account for the intercorrelations of an observed set of variables. In confirmatory factor analysis, on the other hand, one takes a specific hypothesized structure and sees how well it accounts for the observed relationships in the data.

Traditionally, textbooks on factor analysis discuss the topic of exploratory factor analysis at length and in detail, and then they put in something about confirmatory factor analysis in the later chapters. We, however, find it instructive to proceed in the opposite direction, to consider first confirmatory factor analysis and defer an extended treatment of exploratory factor analysis until later. The reason for this is that path models provide a natural and convenient way of representing the hypothesized causal structures involving latent variables which the confirmatory factor analyst wishes to compare to real-world data.

The origins of factor analysis: Charles Spearman and the two-factor theory of intelligence

As it happens, the original form of factor analysis, invented by the British psychologist Charles Spearman shortly after 1900, was more confirmatory than exploratory, in the sense that Spearman had an explicit theory of intellectual performance that he wished to test against data. Spearman did not use a path representation, Wright not yet having invented it, but Fig. 1.13 represents the essentials of Spearman's theory in the form of a path diagram.

Spearman hypothesized that performance on each of a number of intellectual tasks shared something in common with performance on all other intellectual tasks, a factor of general intellectual ability that Spearman called "g." Performance on each task also involved a factor of skills specific to that task, hence the designation "two-factor theory." In Spearman's (1904) words: "All branches of intellectual activity have in common one fundamental function (or group of functions), whereas the remaining or specific elements of the activity seem in every case to be wholly different from that in all the others" (p. 284).

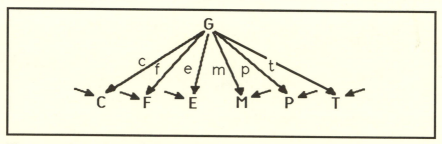

Fig. 1.13 Path representation of Spearman's two-factor theory.

Spearman obtained several measures on a small group of boys at an English preparatory school: a measure of pitch discrimination, a ranking of musical talent, and examination grades in several academic areas-- Classics, French, English studies, and Mathematics. Fig. 1.13 represents his two-factor theory, as applied to these data. The letter G at the top of the figure represents the latent variable of general intellectual ability, C, F, E, and M at the bottom represent observed performances in the academic subjects, P stands for pitch discrimination and T for musical talent. General intellectual ability is assumed to contribute to all these performances. Each also involves specific abilities, represented by the residual arrows.

If Spearman's theory provides an adequate explanation of these data, the path diagram implies that the correlation between any two tasks should be equal to the product of the paths connecting them to the general factor: the correlation between Classics and Mathematics should be *cm*, that between English and French should be *ef*, between French and musical talent *ft*, and so on. Because we are attempting to explain 6x5/2 = 15 different observed correlations by means of 6 inferred values--the path coefficients *c*, *f*, *e*, *m*, *p*, and *t*--a good fit to the data is by no means guaranteed. If one is obtained, it is evidence that the theory under consideration has some explanatory power. Fig. 1.14 gives the correlations for part of Spearman's data: Classics, English, Mathematics, and pitch discrimination.

If the single general-factor model fit the data exactly, we could take the intercorrelations among any three variables and solve for the values of

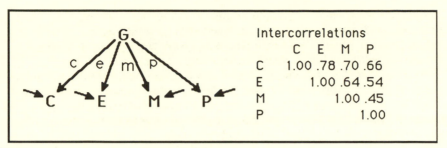

Fig. 1.14 Data to illustrate the method of triads.

the three respective path coefficients, since they would provide three equations in three unknowns. For example:

$$r_{CE}xr_{CM}/r_{EM} = cecm/em = c^2 = .78x.70/.64 = .853; c = .92$$
$$r_{EM}xr_{CE}/r_{CM} = emce/cm = e^2 = .64x.78/.70 = .713; e = .84$$
$$r_{CM}xr_{EM}/r_{CE} = cmem/ce = m^2 = .70x.64/.78 = .574; m = .76.$$

This procedure has been given a name; it is called the *method of triads*. If the data, as here, only approximately fit a model with a single general factor, one will get slightly different values for a particular path coefficient depending on which triads one uses. For example, we may solve for *m* in two other ways from these data:

$$r_{CM}xr_{MP}/r_{CP} = cmmp/cp = m^2 = .70x.45/.66 = .477; m = .69$$
$$r_{EM}xr_{MP}/r_{EP} = emmp/ep = m^2 = .64x.45/.54 = .533; m = .73.$$

These three values of *m* are not very different. One might consider simply averaging them to obtain a compromise value. A slightly preferable method, because less vulnerable to individual aberrant values, adds together the numerators and denominators of the preceding expressions, and then divides:

$$m^2 = \frac{.70x.64 + .70x.45 + .64x.45}{.78 + .66 + .54} = .531; m = .73$$

You may wish to check your understanding of the method by confirming that it yields .97 for *c*, .84 for *e*, and .65 for *p*, for the data of Fig. 1.14.

We may get some sense of how accurately our solution can account for the observed intercorrelations among the four variables, by producing the intercorrelation matrix implied by the paths: i.e, *ce, cm, cp*, etc.

$$\begin{array}{ccc} .81 & .71 & .63 \\ & .61 & .55 \\ & & .47 \end{array}$$

19

As is evident, the implied correlations under the model do not differ much from the observed correlations--the maximum absolute difference is .03. The assumption of a single general factor plus a residual factor for each measure does a reasonable job of accounting for the data.

We may as well go on and estimate the variance accounted for by each of the residual factors. Following the path model, the proportion of the variance of each test accounted for by a factor equals the correlation of that test with itself by way of the factor (the sum of the paths to itself via the factor). In this case these have the value c^2, e^2, etc. The variances due to the general factor are thus .93, .70, .53, and .42 for Classics, English, Mathematics, and pitch discrimination, repectively, and the corresponding residual variances due to specific factors are .07, .30, .47, and .58. In traditional factor analytic terminology, the variance a test shares with other tests in the battery is called its *communality*, symbolized h^2, and the variance not so shared is called its *uniqueness*, symbolized u^2. The communalities of the four measures are thus .93, .70, .53, and .42, and their uniquenesses .07, .30, .47, and .58. Pitch discrimination has the least in common with the other three measures, Classics has the most.

The observant reader will notice that the communality and uniqueness of a variable are just expressions in the factor analytic domain of the general notion of the predicted (R^2) and residual variance of a downstream variable in a path diagram, as discussed earlier in the chapter.

The path coefficients c, e, m, etc. are in factor-analytic writing called the *factor pattern coefficients* (or more simply, the *factor loadings*). The correlations between the tests and the factors, here numerically the same as the pattern coefficients, are collectively known as the *factor structure*.

The centroid method of solving for a general factor

The solution methods we have described can readily be carried out with pencil and hand calculator for small matrices, but the amount of computational labor ascends steeply as the size of the correlation matrix increases. For instance, the four-variable submatrix from Spearman provides 3 triads for each of the four variables, or 12 altogether. The original six-variable matrix yields five times as many, a total of 60 triads.

A more practical way of solving for a single factor underlying a set of intercorrelations was devised by the British factorist Cyril Burt (1917). It is called the centroid method. Table 1-1 illustrates its basic logic.

The square array of paired symbols represents the intercorrelation matrix of the six measures; for example, under the path model of Fig. 1.13, the correlation between Classics and French is cf (or equivalently fc). The main diagonal contains the communalities c^2, f^2, and so on. Now if we take Σ to represent the sum $c + f + e + m + p + t$, the sums of the columns of the table can be given as $c\Sigma$, $f\Sigma$, $e\Sigma$, etc. because each column contains a single term paired in turn with each of the six. The sum of the column totals is then equal to $\Sigma\Sigma$, as

Table 1-1 Logic of the centroid method, applied to Spearman example

---

c^2	fc	ec	mc	pc	tc
cf	f^2	ef	mf	pf	tf
ce	fe	e^2	me	pe	te
cm	fm	em	m^2	pm	tm
cp	fp	ep	mp	p^2	tp
ct	ft	et	mt	pt	t^2

let $\Sigma = c+f+e+m+p+t$

$c\Sigma + f\Sigma + e\Sigma + m\Sigma + p\Sigma + t\Sigma = (c+f+e+m+p+t)\Sigma = \Sigma\Sigma$

$$\Sigma = \sqrt{\Sigma\Sigma}$$

$c = c\Sigma/\Sigma$, $f = f\Sigma/\Sigma$, etc.

---

shown, and the square root Σ of this quantity can be divided into each of the column totals to solve for c, f, e, m, p, and t, the desired paths.

This method can readily be applied to an intercorrelation matrix, regardless of size, to obtain the path coefficients for a general factor: One simply sums the columns, and divides each by the square root of the sum of these sums. The only catch is, how does one get the communalities to put in the diagonal of the matrix? The answer is that even a rough estimate will do, especially if the correlation matrix is fairly large. We have more to say in a later chapter about different methods of communality estimation. For the moment, one method we can use is to estimate the communality from two or three triads involving the variable.

Table 1-2 applies the centroid method to the full Spearman example. For C, E, M, and P we have used the communalities estimated earlier from the four-variable set (3 triads each). For F and T we have arbitrarily used the three triads obtained by pairing F and T with C,E, E,M, and M,P. The total of the row totals is 22.19. Dividing the row totals by the square root of this, 4.7106, yields the path coefficients as shown, and squaring them the communalities. Note that the path coefficients for Classics, English, Mathematics, and pitch discrimination--.96, .82, .74, and .66--do not differ much from the .97, .84, .73, and .65 obtained earlier by the method of triads with the partial data set.

The obtained communalities also do not differ much from the estimated ones--the maximum discrepancy is .04 for E. If we wanted to see how much difference this makes (and incidentally, get slightly improved values for the path coefficients), we could put these obtained communality values in place of the estimates we used originally and repeat the calculations leading to the path coefficients. We do not bother to show the calculations. In no case is the resulting difference greater than one in the second decimal place.

The bottom of Table 1-2 shows the correlations implied by the path coefficients. Again, we see that they do a reasonably good--but not a perfect--job of reproducing the original correlations. About half the values are within one or two points of the original in the second decimal place; the largest

Table 1-2 Numerical illustration of the centroid method, applied to the Spearman example

	C	F	E	M	P	T	
C	(.93)	.83	.78	.70	.66	.63	
F	.83	(.77)	.67	.67	.65	.57	
E	.78	.67	(.70)	.64	.54	.51	
M	.70	.67	.64	(.53)	.45	.51	
P	.66	.65	.54	.45	(.42)	.40	
T	.63	.57	.51	.51	.40	(.42)	
	4.53	4.16	3.84	3.50	3.12	3.04	ΣΣ=22.19
							Σ=4.7106
	.962	.883	.815	.743	.662	.645	path coeff.
	.92	.78	.66	.55	.44	.42	h^2

	C	F	E	M	P	T
C		.85	.78	.71	.64	.62
F	-.02		.72	.66	.58	.57
E	.00	-.05		.61	.54	.53
M	-.01	.01	.03		.49	.48
P	.02	.07	.00	-.04		.43
T	.01	.00	-.02	.03	-.03	

Note: Upper matrix--observed intercorrelations; estimated communalities in diagonal. Lower matrix--above diagonal, correlations implied by path coefficients; below diagonal, residual correlations unaccounted for by general factor.

discrepancy is .07. We return later to the question of just how good a fit is "good enough."

One final point: The centroid method assumes that all variables are scored in the same direction with respect to the general factor, so that their intercorrelations are predominantly positive. Suppose that Spearman had happened to use a measure of pitch discrimination such as errors or latency that was scored so that a high score meant that discrimination was poor. Its correlation with all the other measures would then be negative, and this would cause trouble. In such a case one could reverse the scoring of the original measure (say, by subtracting each score from the highest possible score plus 1), or, equivalently, simply reverse all the signs of the correlations in the matrix that involve that variable. This is known as *reflecting* the variable and should be carried out if necessary before applying the centroid method. Note that if one were to fail to do this, it would not affect the column total for the variable in question, other than changing its sign, but it would decrease the totals for all the other variables and hence distort the analysis.

More than one common factor

As soon became evident to Spearman's followers and critics, not all observed sets of intercorrelations are well explained by a model containing only one general factor; factor analysts soon moved to models in which more than one latent variable was postulated to account for the observed intercorrelations among measures. Such latent variables came to be called *common* factors, rather than *general* factors because, although they were common to several of the variables under consideration, they were not general to all. There remained, of course, *specific* factors unique to each measure. Fig. 1.15 gives

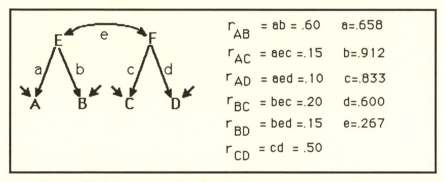

r_{AB} = ab = .60	a=.658
r_{AC} = aec =.15	b=.912
r_{AD} = aed =.10	c=.833
r_{BC} = bec =.20	d=.600
r_{BD} = bed =.15	e=.267
r_{CD} = cd = .50	

Fig. 1.15 A simple factor model with two factors.

an example of a path diagram in which there are two latent variables, E and F, and four observed variables, A, B, C, and D. E is hypothesized as influencing A and B, and F as influencing C and D. E and F are correlated. (These are, of course, arbitrary assumptions--one could have postulated that E influenced C or D, or that F did not affect C, or that E was uncorrelated with F.)

In the path diagram of Fig. 1.15 there are five unknowns, the paths *a, b, c,* and *d,* and the correlation *e* between the two latent variables. There are six equations, shown to the right of the diagram, based on the six intercorrelations between pairs of observed variables. A set of hypothetical values of observed correlations is given to the right of the second equals sign. Because there are more equations than unknowns, one might expect that a single exact solution would not be available, and indeed this is the case. An iterative least squares solution, carried out in a manner to be discussed in the next chapter, yielded the values shown to the far right of Fig. 1.15. Table 1-3 reports a factor analysis solution derived from these values. The factor pattern represents the values of the paths from factors to variables; i.e., the paths *a* and *b* and two zero paths from E to A, B, C, and D, and the corresponding paths from F. The factor structure presents the correlations of the variables with the factors: for factor E these have the values *a, b, ce,* and *de,* respectively, and for factor F, *ae, be, c,* and *d.* The communalities (h^2) are in this case simply a^2, b^2, c^2, and d^2,

Table 1-3 Factor solution for the two-factor problem of Fig. 1.15

Variable	Factor pattern		Factor structure		h^2
	E	F	E	F	
A	.66	.00	.66	.18	.43
B	.91	.00	.91	.24	.83
C	.00	.83	.22	.83	.69
D	.00	.60	.16	.60	.36

Factor intercorrelations

	E	F
E	1.00	.27
F	.27	1.00

Reproduced and residual correlations

	A	B	C	D
A		.600	.146	.105
B	.000		.203	.146
C	.004	-.003		.500
D	-.005	.004	.000	

because each variable is influenced by only one factor. Finally, the correlation between E and F is just *e*.

The reproduced correlations (those implied by the path values) and the residual correlations (the differences between observed and implied correlations) are shown at the bottom of Table 1-3. The reproduced correlations are obtained by inserting the solved values of *a, b, c,* etc. into the equations of Fig. 1.15: $r_{AB} = .658 \times .912$, $r_{AC} = .658 \times .267 \times .833$, and so on.

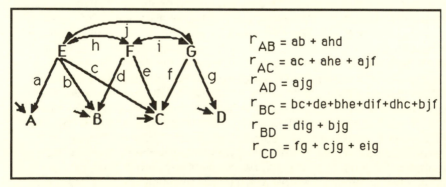

$$r_{AB} = ab + ahd$$
$$r_{AC} = ac + ahe + ajf$$
$$r_{AD} = ajg$$
$$r_{BC} = bc+de+bhe+dif+dhc+bjf$$
$$r_{BD} = dig + bjg$$
$$r_{CD} = fg + cjg + eig$$

Fig. 1.16 A more complex three-factor model.

A more complex model with three factors is shown in Fig. 1.16. Because this model has 10 unknowns and only 6 equations, it is underdetermined and cannot be solved as it stands. However, if one were to fix sufficient values by a priori knowledge or assumption, one could solve for the remaining values. The factor solution in symbolic form is given in Table 1-4. By inserting the known and solved-for values in place of the unknowns, one could obtain numerical values for the factor pattern, the factor structure, the communalities, and the factor intercorrelations. Also, one could use the path equations of Fig. 1.16 to obtain the implied correlations and thence the residuals.

Notice that the factor pattern is quite simple in terms of the paths, but that the factor structure (the correlations of factors with variables) and the communalities are more complex functions of the paths and factor intercorrelations.

Table 1-4 Factor solution of Fig. 1.16, in symbolic form

Variable	Factor pattern			Factor structure			h^2
	E	F	G	E	F	G	
A	a	0	0	a	ha	ja	a^2
B	b	d	0	b+hd	d+hb	id+jb	b^2+d^2+2bhd
C	c	e	f	c+he	e+hc	f+ie	$c^2+e^2+f^2+2che$
				+jf	+if	+jc	+2eif+2cjf
D	0	0	g	jg	ig	g	g^2

Factor intercorrelations

	E	F	G
E	1.0	h	j
F	h	1.0	i
g	j	i	1.0

Structural Equations

An alternative way of representing a path diagram is as a set of *structural equations*. Each equation expresses a downstream variable as a function of the causal paths leading into it. There will be as many equations as there are downstream variables. Fig. 1.17 shows one of the path diagrams considered earlier. It has one downstream variable, hence one structural equation: The score of a person on variable C is an additive function of his scores on A, B, and X. If the variables are obtained in standard-score form for a set of subjects, the values of the weights *a, b,* and *d* required to give a best fit to the data in a least squares sense turn out to be just the standardized partial regression coefficients, or path coefficients, we have discussed earlier.

25

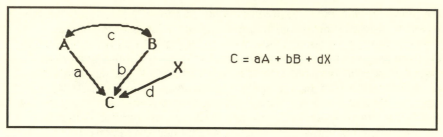

Fig. 1.17 A structural equation based on a path diagram

Figure 1.18 gives a slightly more complex example, based on the earlier Fig. 1.3. Now there are two downstream variables, A_2 and B_3. A_2 can be expressed as a weighted additive function of the three source variables A_1, B_1,

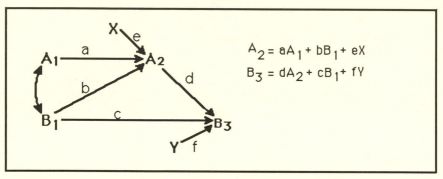

Fig. 1.18 Structural equations based on the path diagram of Fig. 1.3.

and X, as shown in the first equation, whereas B_3 can be expressed in terms of A_2, B_1, and Y. Note that to construct a structural equation one simply includes a term for every straight arrow leading into the downstream variable. The term consists of the variable at the tail of the arrow times the path coefficient associated with it.

For a final example, consider the factor analysis model of Fig. 1.16 in the preceding section. The structural equations are as follows (X_A, X_B, etc. represent the terms involving the residual arrows):

$$A = aE + X_A$$
$$B = bE + dF + X_B$$
$$C = cE + eF + fG + X_C$$
$$D = gG + X_D$$

26

Notice that the equations are closely related to the rows of the factor pattern matrix (Table 1-4) with residual terms added. The solution of the set of structural equations corresponds essentially to the solution for the paths in the path diagram and would be similarly underdetermined in this instance. Again, by previously defining a sufficient number of the unknowns, the equations could be solved for those remaining.

The structural equation approach to causal models originated in economics and the path approach in biology. The preference for one or the other in the various branches of the social and behavioral sciences is likely to be as much a matter of historical accident as anything else. Because the two, so far as we are concerned, are essentially just different ways of representing the same facts, use of one or the other approach, for someone conversant with both, can be a matter of personal preference or of the audience to whom one is communicating.

Equivalence of path and structural equation representation

Some readers may prefer to take the equivalence of path and structural equation representation on faith, but if you are skeptical, this equivalence can be seen via fairly straightforward algebra, using an example from Fig. 1.18 in the preceding section.

By path-tracing rules, the correlation between variables A_2 and B_3 in Fig. 1.18 is:

$$r_{A2B3} = d + bc + a\ r_{A1B1}c .$$

Assuming that all variables are measured in standard scores, we may write the correlation between A_2 and B_3 as $\Sigma A_2 B_3 / N$. Substituting the structural equations for A_2 and B_3 from Fig. 1.18, we have:

$$r_{A2B3} = \Sigma A_2 B_3 / N = [\Sigma(aA_1 + bB_1 + eX)(dA_2 + cB_1 + fY)]/N .$$

Multiplying out within the square brackets, we obtain for the righthand side:

$$[\Sigma(adA_1 A_2 + acA_1 B_1 + afA_1 Y + bdB_1 A_2 + bcB_1{}^2 + bfB_1 Y + edXA_2 + ecXB_1 + efXY)]/N.$$

Separating the terms in the summation, and taking constants outside the summation signs yields:

$$ad\Sigma A_1 A_2/N + ac\Sigma A_1 B_1/N + af\Sigma A_1 Y/N + bd\Sigma B_1 A_2/N + bc\Sigma B_1{}^2/N +$$
$$bf\Sigma B_1 Y/N + ed\Sigma XA_2/N + ec\Sigma XB_1/N + ef\Sigma XY/N.$$

Note that we now have an expression in terms of variances and covariances, which we can simplify further because with standard scores variances equal unity and covariances are correlations. Furthermore, the assumption that residuals are uncorrelated allows us to drop all but one of the covariance terms involving X or Y (the exception is the covariance of X and A_2, which is *e*). Thus:

$$r_{A2B3} = ad \ r_{A1A2} + ac \ r_{A1B1} + bd \ r_{B1A2} + bc + e^2 d.$$

Now, the second and fourth terms on the right-hand side correspond directly to terms in the path expression, so it remains to show that what is left from the two expressions is equal, i.e.:

$$ad \ r_{A1A2} + bd \ r_{B1A2} + e^2 d = d .$$

The left side of this equation may be written:

$$d(a \ r_{A1A2} + b \ r_{B1A2} + e^2),$$

and, as it turns out, the expression in parentheses equals the variance of A_2, which is 1.0. To see this, note that r_{A1A2} may be written by path rules as $a + r_{A1B1} b$ and that r_{B1A2} may be written as $b + r_{A1B1} a$, so that the whole expression in parentheses equals:

$$(a^2 + b^2 + e^2 + 2ab \ r_{A1B1}),$$

which by path-tracing rules is the variance of A_2.

Thus, the path and the structural equation representations are equivalent in this instance. You may wish to satisfy yourself of their correspondence in some other cases.

Original and Standardized Variables

So far, we have assumed we were dealing throughout with standardized variables. This has simplified the presentation, but is not a necessary restriction. Path, factor, and structural equation analyses can be carried out with variables in their original scale units as well as with standardized variables. In practice, structural equation analysis is usually done in rawscore units, path analysis is done both ways, and factor analysis is usually done with standardized variables. But this is often simply a matter of tradition or (what amounts to much the same thing) of the input-output practices of the particular computer program used. There are occasions on which the standardized and rawscore approach each has definite advantages, so it is important to know that one can convert the results of one to the other form and be able to do so when the occasion arises.

Another way of making the distinction between analyses based on standardized and raw units is to say that in the first case one is analyzing correlations, and in the second, covariances. In the first case one decomposes a correlation matrix among observed variables into additive components; in the second case one so decomposes a variance-covariance matrix. The curved arrows in a path diagram are correlations in the first case, covariances in the second. In the first case a straight arrow in a path diagram stands for a standardized partial regression coefficient, in the second case for a rawscore partial regression coefficient. In the first case a .5 beside a straight arrow leading from years of education to annual income means that, other things equal, people in this particular population who are one standard deviation above the mean in education tend to be half a standard deviation above the mean in income. In the second case, if education is measured in years and income in dollars, a 5,000 alongside the straight arrow between them means that, other things equal, an increase of 1 year in education represents an increase of $5,000 in annual income (in this case, .5 would mean 50 cents!). In each case the arrow between A and B refers to how much change in B results from a given change in A, but in the first case change is measured in standard deviation units of the two variables, and in the second case, in the ratio of their rawscore units (dollars of income per year of education).

Standardized regression coefficients are particularly useful when comparisons are to be made across different variables, unstandardized regression coefficients when comparisons are to be made across different populations.

In the first case, it is difficult to compare the relative importance of education and occupational status in influencing income if the respective rawscore coefficients are 5,000 and 300, based on income in dollars, education in years, and occupational status on a 100-point scale. But if the standardized regression coefficients are .5 and .7, respectively, the greater relative influence of occupational status is immediately obvious.

In the second case, rawscore regression coefficients have the merit of independence of the particular ranges of the two variables involved in the study. If one study happens to have sampled twice as great a range of education as another, a difference in years of education that is, say, one-half a standard deviation in the first study would be a full standard deviation in the second. A standardized regression coefficient of .3 in one study would then describe exactly the same effect of education on income as a standardized regression coefficient of .6 in the other. This is a confusing state of affairs at best and could be seriously misleading if the reader is unaware of the sampling difference between the studies. A rawscore regression coefficient of $2,000 per year of education would, however, have the same meaning across the two studies. If the relevant standard deviations are known, a correlation can readily be transformed into a covariance, or vice versa, or a rawscore into a standardized regression coefficient and back, allowing one freely to report results in either or both ways, or to carry out calculations in one mode and report them in the other, if desired. (We qualify this statement later-- calculational methods can sometimes be sensitive to the scale on which variables are expressed--but it will do for now.)

The relationships between covariances and correlations are simple:

$$cov_{12} = r_{12} \, s_1 s_2$$
$$r_{12} = cov_{12}/s_1 s_2,$$

where cov_{12} stands for the covariance between variables 1 and 2, r_{12} for the correlation between them, and s_1 and s_2 for their respective standard deviations.

The relationships between rawscore and standardized path coefficients are equally simple. To convert a standardized path coefficient to its rawscore form, *multiply it by the ratio of the standard deviations of its head to its tail variable.* To convert a rawscore path coefficient to standardized form, invert the process: Multiply by the ratio of the standard deviations of its tail to its head variable.

These rules generalize to a series of path coefficients, as illustrated by Fig. 1.19 and Table 1-5. The first line in the table shows, via a process of substituting definitions and cancelling, that the series of rawscore path coefficients *a*b*c** is equal to the series *abc* of standardized path coefficients multiplied by the ratio of standard deviations of its head and tail variables. The second line demonstrates, by substituting the result from the preceding line, the converse transformation from rawscore to standardized coefficients.

The last line in Table 1-5 shows that the covariance between A and D may be obtained as a product of the variance of A, the topmost variable in the path, and the series of rawscore coefficients *a*b*c** leading to D. First, the definition for covariance is substituted, then *abc* is substituted for r_{AD}, then the

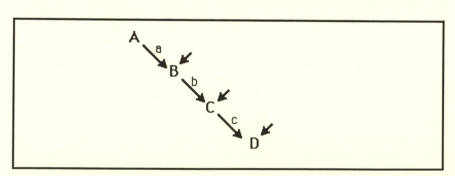

Fig. 1.19 Path diagram to illustrate rawscore and standardized path coefficients.

Table 1-5 Transformation of a sequence of paths from rawscore to standardized form (example of Fig. 1.19)

$a^* \ b^* \ c^* = a(s_B/s_A) \ b(s_C/s_B) \ c(s_D/s_C) = abc(s_D/s_A)$

$a^* \ b^* \ c^*(s_A/s_D) = abc(s_D/s_A)(s_A/s_D) = abc$

$cov_{AD} = r_{AD}s_A s_D = abc \ s_A s_D = s_A^2 a^* \ b^* \ c^*$

Note: Asterisks designate rawscore path coefficients.

expression for *abc* from line 2.

This illustrates a general rule for expressing the value of a compound path between two variables in terms of path coefficents (stated for a vertically oriented diagram):

The value of a compound path between two variables is equal to the product of the rawscore path coefficients and the topmost variance or covariance in the path.

The tracing of compound paths according to Wright's rules, and adding compound paths together to yield the overall covariance, proceed in just the same way with rawscore as with standardized coefficients. The covariance between two variables in the diagram is equal to the sum of the compound paths between them. If there is just a single path between two variables, the covariance is equal to the value of that path. The two path diagrams in Fig. 1.20 illustrate the rule for compound paths headed by a variance and a covariance respectively.

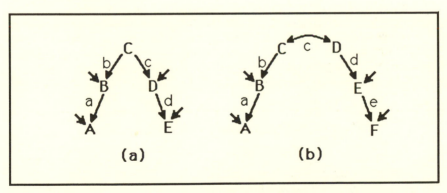

(a)   (b)

Fig. 1.20 Rawscore paths with (a) a variance and (b) a covariance.

The equivalence to a covariance between the terminal variables is shown in Table 1-6. In the first row of the table, the rule is stated for rawscore path coefficients; in the second row these are transformed to standardized coefficients (note that the path *c* on the right is r_{CD}, and the covariance is redefined as a correlation times two standard deviations); in the third row

Table 1-6 Demonstrations of rawscore compound path rules, for path diagrams of Fig. 1.20

(a)	(b)
$cov_{AE} = a^* b^* s_C^2 c^* d^*$	$cov_{AF} = a^* b^* cov_{CD} d^* e^*$
$\quad = ab(s_A/s_C) s_C^2 cd(s_E/s_C)$	$\quad = ab(s_A/s_C)c\ s_C s_D de(s_F/s_D)$
$\quad = abcd\ s_A\ s_E$	$\quad = abcde\ s_A\ s_F$
$\quad = r_{AE}\ s_A\ s_E$	$\quad = r_{AF}\ s_A\ s_F$

---

redundant terms are cancelled; and in the fourth the standardized path sequence is reexpressed as a correlation to yield the definition of a covariance.

Notice that the rule for evaluating compound paths when using rawscore path coefficients is different from that for standardized coefficients only by the inclusion of one variance or covariance in each path product. Indeed, one can think of the standardized rule as a special case of the rawscore rule, because the variance of a standardized variable is 1, and the covariance between standardized variables is just the correlation coefficient.

Because the rawscore expressions tend to be a bit clumsier to work with, a reasonable way to proceed in many cases is to carry out the analysis using standardized variables, and then to convert path coefficients between measured variables to rawscore form, using the standard deviations of the variables. Both standardized and rawscore coefficients can then be reported. Path coefficients involving latent variables, for which standard deviations will not be available, can be left in their standardized form. An alternative approach, fairly common among those who prefer to work with covariances and rawscore coefficients, is to assign an arbitrary value, usually 1.0, to a path linking the latent variable to an observed variable, thereby implicitly expressing the latent variable in units based on the observed variable. Several examples of this procedure appear in later chapters.

Differences From Some Related Topics

We need also to be clear about what this book does *not* cover. In this section some related topics, which might easily be confused with latent variable analysis as we discuss it, are distinguished from it.

Manifest versus latent variable models

Many multivariate statistical methods, including some of those most familiar to social and behavioral scientists, do not involve latent variables. Instead, they deal solely with linear composites of *observed* variables. In ordinary multiple regression, for example, one seeks for an optimally weighted composite of measured independent variables to predict an observed dependent or criterion variable. In discriminant analysis, one seeks composites of measured variables that will optimally distinguish among members of specified groups. In canonical analysis one seeks composites that will maximize correlation across two sets of measured variables.

Path and structural analysis come in both forms: all variables measured or some not. Many of the earlier applications of such methods in economics and sociology were confined to manifest variables. The effort was to fit causal models in situations where all the variables involved were observed. Biology and psychology, dealing with events within the organism, tended to place an earlier emphasis on the latent variable versions of path analysis. As researchers in all the social sciences become increasingly aware of the distorting effects of measurement errors on causal inferences, latent variable methods have increased in popularity, especially in theoretical contexts. In applied situations, where the practitioner must work with existing measures, errors and all, the manifest variable methods retain much of their preeminence.

Factor analysis is usually *defined* as a latent variable method--the factors are unobserved hypothetical variables that underlie and explain the observed correlations. The corresponding manifest variable method is called *component analysis* --or, in its commonest form, the method of *principal components.* Principal components are linear composites of observed variables; the factors of factor analysis are always inferred entities, whose nature is at best consistent with a given set of observations, never entirely determined by them.

In general, statistical theory is better developed for manifest than for latent variable methods, although there has been a good deal of progress in the latter area in recent years.

Single versus multiple latent variables

A good deal of recent interest among psychometricians has centered on *item reponse theory*, sometimes called *latent trait theory,* in which a latent variable is fit to responses to a series of test items. We do not discuss these methods in this book. They typically focus on fitting a single latent variable (the underlying trait being measured) to the responses of subjects to a set of test items, whereas our principal concern is with fitting models involving multiple latent variables.

More critically, however, the relationships dealt with in item response theory are typically nonlinear: Two- or three-parameter latent curves are fitted, such as the logistic. This book is primarily concerned with methods that assume linear relationships, although we briefly consider some extensions to nonlinear cases in the final chapter.

Latent classes versus latent dimensions

Another substantial topic which this book does not attempt to cover is the modeling of latent classes or categories underlying observed relationships. This topic is often called, for historical reasons, *latent structure analysis* (Lazarsfeld, 1950), although the more restrictive designation *latent class analysis* better avoids confusion with the latent variable methods described in this book. The methods we discuss also are concerned with "latent structure," but it is structure based on relations among continuous variables rather than on the existence of discrete underlying categories.

Chapter 1 Notes

Path analysis. An introductory account, somewhat oriented toward genetics, is Li (1975). The statement of Wright's rules in this chapter is adapted from Li's. Kenny (1979) provides another introductory presentation with a slightly different version of the path-tracing rules: A single rule--a variable entered via an arrowhead cannot be left via an arrowhead--covers rules 2 and 3. The sociologist O. D. Duncan (1966) is usually credited with rediscovering path analysis for social scientists; Werts and Linn (1970) wrote a paper calling psychologists' attention to the method.

Factor analysis. Maxwell (1977) has a brief account of some of the early history. Mulaik (1986) brings matters up to the present. Gorsuch (1983) provides an up-to-date and readable general treatment of factor analysis. For an explicit distinction between the exploratory and confirmatory varieties, see Jöreskog and Lawley (1968).

Structural equations. These come from econometrics--for some relationships between econometrics and psychometrics, see Goldberger (1971) and a special issue of the journal *Econometrics* edited by de Leeuw, Keller, and Wansbeek (1983). An introduction to structural equation models for social scientists is provided in a book by Duncan (1975). Texts by Heise (1975), James, Mulaik, and Brett (1982), Dwyer (1983), and Long (1983a,b) also provide coverage of path and structural analysis. A historical perspective is presented by Bentler (1986).

Direct and indirect effects. For a discussion of such effects, and the development of matrix methods for their systematic calculation, see Fox (1980, 1985).

Under and overdetermination in path diagrams. This is discussed as *identification* in the structural equation literature. It is not a simple matter: For an example see Bollen and Jöreskog (1985). As one complication, models may sometimes be identified for some values of their unknown paths, and not for others (Rindskopf, 1984a). In simple cases, an algebraic analysis of the equations can ascertain whether or not all the unknowns are determinate.

Jöreskog and Sörbom (1984, I.23-24) suggest some empirical strategies for dealing with more complex cases, and Land and Felson (1978) discuss some approaches for investigating the sensitivity of one's results to arbitrary assumptions made in order to achieve identification.

Original and standardized variables. Their relative merits are debated by Tukey (1954) and Wright (1960). For a recent discussion, see Kim and Ferree (1981).

Related topics. Several examples of manifest-variable path and structural analysis may be found in Marsden (1981), especially Part II. Principal component analysis is discussed in most factor analysis texts (e.g., Harman, 1976). A number of papers on item response theory comprise a special issue of *Applied Psychological Measurement* (Hambleton & van der Linden, 1982). The basic treatment of latent class analysis is Lazarsfeld and Henry (1968). For a broad, general treatment of structural models that covers both quantitative and qualitative variables, see Kiiveri and Speed (1982).

Journal sources. Some journals that frequently publish articles on developments in the area of latent variable models include *Psychometrika, Sociological Methods and Research, Multivariate Behavioral Research , The British Journal of Mathematical and Statistical Psychology,* and *Applied Psychological Measurement.* See also the annual series, *Sociological Methodology.*

Chapter 1 Exercises

Note: Answers to exercises are in the back of the book, preceding the References.

1. Draw a path diagram of the relationships among impulsivity and hostility at time 1 and delinquency at time 2, assuming that the first two influence the third but not vice versa.

2. Draw a path diagram of the relationships among ability, motivation, and performance, each measured on two occasions.

Questions 3 to 9 refer to the path diagram in Fig. 1.21 (assume all variables are in standard-score form unless otherwise specified).

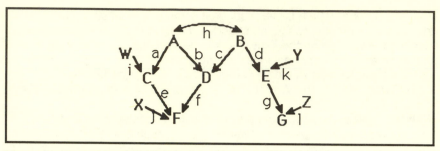

Fig. 1.21 Path diagram for problems 3 to 9.

3. Identify the source and downstream variables in the diagram.

4. What assumption is made about the causation of variable D?

5. Write path equations for the correlations r_{AF}, r_{DG}, r_{CE}, and r_{EF}.

6. Write path equations for the variances of C, D, and F.

7. If variables A, B, F, and G are measured, would you expect the path diagram to be solvable? (Explain why or why not.)

8. Now, assume that the variables in Fig. 1.21 are *not* standardized. Write path equations, using rawscore coefficients, for the covariances c_{CD}, c_{FG}, c_{AG} and the variances s_G^2 and s_D^2.

9. Write structural equations for the variables D, E, and F in Fig. 1.21.

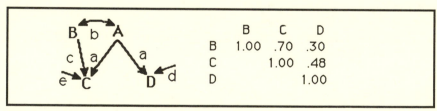

	B	C	D
B	1.00	.70	.30
C		1.00	.48
D			1.00

Fig. 1.22 Path diagram for problem 10.

10. Given the path diagram shown in Fig. 1.22 and the observed correlations given to the right, solve for *a, b, c, d,* and *e.*

11. The following intercorrelations among three variables are observed:

	A	B	C
A	1.00	.42	.12
B		1.00	.14
C			1.00

Solve for the loadings on a single common factor, using the method of triads.

12. Four putative measures of creativity, tests W, X, Y, and Z, are given to a large group of high school students. The intercorrelations among the tests follow:

	W	X	Y	Z
W	1.00	.40	.50	.30
X		1.00	.55	.35
Y			1.00	.40
Z				1.00

Calculate the factor loadings using the centroid method--use an estimate of the communality for each variable based on a triad with its two nearest neighbors in the matrix.

13. Show the equivalence of path rules and structural equation algebra for the covariance between A_2 and B_3 in Fig. 1.18 for the *un*standardized case. (Assume that all variables are in deviation score form.)

Chapter Two:
Fitting Path Models

In this chapter we consider the processes used in actually fitting path models to data on a realistic scale, and evaluating their goodness of fit. This implies computer-oriented methods. This chapter is somewhat more technical than Chapter 1. Some readers on a first pass through the book might prefer to read carefully only the section on hierarchical χ^2 tests (pp. 62-67) and then go on to Chapters 3 and 4, coming back to Chapter 2 afterwards. (You need additional Chapter 2 material to do the exercises in Chapters 3 and 4.)

Iterative Solution of Path Equations

In simple path diagrams like those we have considered so far, direct algebraic solution of the set of implied equations is often quite practicable. But as the number of observed variables goes up, the number of intercorrelations among them, and hence the number of equations to be solved, increases rapidly. (The number of equations is given by $n(n-1)/2$, where n is the number of observed variables.) Furthermore, path equations by their nature involve product terms, because a compound path is the product of its component arrows. Product terms make the equations recalcitrant to straightforward matrix procedures that can be used to solve sets of linear simultaneous equations. As a result of this, large sets of path equations are in practice usually solved by iterative (= repetitive) trial-and-error procedures, carried out by computers.

    The general idea is simple. An arbitrary set of initial values of the paths serves as a starting point. The correlations or covariances implied by these values are calculated and compared to the observed values. Because the initial values are arbitrary, the fit is likely to be poor. So one or more of the initial trial values is changed in a direction that improves the fit, and the process is repeated with this new set of trial values. This cycle is repeated again and again, each time modifying the set of trial values to improve the agreement between the implied and the observed correlations. Eventually, a set of values

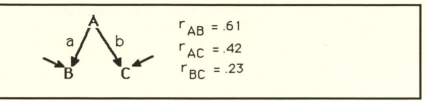

$r_{AB} = .61$

$r_{AC} = .42$

$r_{BC} = .23$

Fig. 2.1 A simple path diagram illustrating an iterative solution.

will be reached that cannot be improved on--the process, as the numerical analysts say, has "converged" on a solution. If all has gone well, this will be the optimum solution that is sought.

Let us illustrate this procedure with the example shown in Fig. 2.1. A simple case like this one might be solved in more direct ways, but we use it to demonstrate an iterative solution, as shown in Table 2-1.

We begin in cycle 1 by setting arbitrary trial values of *a* and *b*--for the

Table 2-1 An iterative solution of the path diagram of Fig. 2.1

		Trial values		Correlations			Criterion
		a	b	r_{AB}	r_{AC}	r_{BC}	Σd^2
Observed				.61	.42	.23	
Cycle	1	.5	.5	.50	.50	.25	.018900
	1a	.501	.5	.501	.50	.2505	.018701*
	1b	.5	.501	.50	.501	.2505	.019081
	2	.6	.5	.60	.50	.30	.011400
	2a	.601	.5	.601	.50	.3005	.011451
	2b	.6	.501	.60	.501	.3006	.011645*
	3	.6	.4	.60	.40	.24	.000600
	3a	.601	.4	.601	.40	.2404	.000589
	3b	.6	.401	.60	.401	.2406	.000573*
	4	.6	.41	.60	.41	.246	.000456
	4a	.601	.41	.601	.41	.2464	.000450*
	4b	.6	.411	.60	.411	.2466	.000457
	(5)	.61	.41	.61	.41	.2501	.000504
	5	.601	.41	.601	.41	.2464	.0004503
	5a	.602	.41	.602	.41	.2468	.0004469*
	5b	.601	.411	.601	.411	.2470	.0004514
	6	.602	.41	.602	.41	.2468	.0004469
	6a	.603	.41	.603	.41	.2472	.0004459
	6b	.602	.411	.602	.411	.2474	.0004485*
	(7)	.603	.409	.603	.409	.2462	.0004480

*greater change

39

example we have set each to .5. Then we calculate the values of the correlations r_{AB}, r_{AC}, and r_{BC} that are implied by these path values: they are .50, .50, and .25, respectively. We choose some reasonable criterion of the discrepancy between these and the observed correlations--say, the sum of the squared differences between the corresponding values. In this case this sum is $.11^2 + (-.08)^2 + (-.02)^2$, or .0189.

Next, in steps 1a and 1b, we change each trial value by some small amount (we have used an increase of .001) to see what effect this has on the criterion. Increasing *a* makes things better and increasing *b* makes things worse, suggesting that either an increase in *a* or a decrease in *b* should improve the fit. (If you happen to be familiar with differential calculus, you will recognize that what we are doing is obtaining a rough empirical assessment of the first derivatives of the criterion function with respect to the trial values.) Because the change 1a makes a bigger difference than the change 1b does, suggesting that the criterion will improve faster with a change in *a*, we increase the trial value by 1 in the first decimal place to obtain the new set of trial values in 2. Repeating the process, in lines 2a and 2b, we find that a change in *b* now has the greater effect; the desirable change is a decrease.

Decreasing *b* by 1 in the first decimal place gives the line 3 trial values .6 and .4. In line 3a and 3b we find that increasing either would be beneficial, *b* more so. But increasing *b* in the first decimal place would just undo our last step, yielding no improvement, so we shift to making changes in the second place. (This is not necessarily the numerically most efficient way to proceed, but it will get us there.) Line 4 confirms that the new trial values of .6 and .41 do constitute an improvement. Testing these values in lines 4a and 4b, we find that an increase in *a* is suggested. We try increasing *a* in the second decimal place, but this is not an improvement, so we shift to an increase in the third decimal place (line 5). The tests in lines 5a and 5b suggest that a further increase to .602 would be justified, so we use that in line 6. Now it appears that decreasing *b* might be the best thing to do, line (7), but it isn't an improvement. Rather than go on to still smaller changes, we elect to quit at this point, reasonably confident of at least 2-place precision in our answer of .602 and .410 in line 6 (or, slightly better, the .603 and .410 in 6a).

Now, doing this by hand for even two unknowns is fairly tedious, but it is just the kind of repetitious, mechanical process that computers are good at, and many general and special-purpose computer programs exist that can carry out such minimizations. If you were using a typical general-purpose minimization program, you would be expected to supply it with an initial set of trial values of the unknowns, and a subroutine that calculates the function to be minimized, given a set of trial values. That is, you would program a subroutine that will calculate the implied correlations, subtract them from the observed correlations, and sum the squares of the differences between the two. The minimization program will then proceed to adjust the trial values iteratively, in some such fashion as that portrayed in Table 2-1, until an unimprovable minimum value is reached.

A simple computer implementation of iterative path solutions-- IPSOL

The program IPSOL, given in Appendix B, is a simple version of such a computer program written primarily for didactic purposes. It carries out an iterative model-fitting process identical to that we have just done by hand in Table 2-1. IPSOL, which is short for "iterative path solution," is written in the computer language FORTRAN; for readers not familiar with this language, asterisked comment lines have been supplied to provide a general sense of what is going on.

The program consists of three parts. First, there is an executive routine, which reads in problem specifications and data, calls on the minimization routine to do its work, and prints out the results at the end. Next, there is the minimization routine MINIM itself, which makes the trial-and-error changes in the values of the unknown parameters, starting with relatively large steps and shifting to smaller ones as the solution is approached. It evaluates the effect of proposed changes by calling on the third part of the program, the subroutine FUNCT that calculates the implied values and assesses their fit to the observed ones--in this case, as the sum of the squared discrepancies between the two.

Table 2-2 shows the path equations and sample input for solving the path problem of Table 2-1 via IPSOL. The equations, written in the symbolism of the FORTRAN language, express each of the implied, or calculated values, the CV(i)s on the left, in terms of the unknowns, the X(i)s on the right. The equations correspond to the path diagram of Fig. 2.1: The first says that the first correlation (r_{AB}) is equal to the first unknown (a); the second says that the second correlation (r_{AC}) is equal to the second unknown; and the third says

Table 2-2 Path equations and input for solving sample problem of Fig. 2.1 and Table 2-1 via IPSOL

Path equations for subroutine FUNCT:
$$CV(1) = X(1)$$
$$CV(2) = X(2)$$
$$CV(3) = X(1)*X(2)$$

Input for sample problem:

```
TRIAL RUN OF IPSOL WITH FIG 2.1 PROBLEM
    3  2  2
       .61        .42        .23
    .5 .5
    .2 .2
```

that the third correlation (r_{BC}) is equal to the product of the two unknowns (FORTRAN uses an asterisk to signify multiplication).

The input shown in Table 2-2 supplies the program with the remaining information necessary for carrying out the solution. The first line is a title, which is simply reproduced on the output. The second line gives the problem specifications: 3 observed values, 2 unknowns to be solved for, and 2 runs with different starting values to be carried out. The spacing is important: Each number is right-justified in a 3-digit field (i.e., a single- digit number is preceded by two blanks; a two-digit number would be preceded by one blank).

The third line of the input provides the observed values to be fitted, the correlations .61, .42, and .23 (note that they must be in the same order as the equations calculating the corresponding implied values). An 8-digit field is provided for each (to allow space for entering covariances) with a decimal point required. (The location of the number within the 8-digit field is arbitrary.) Ten observed values can be entered per line, in as many lines as necessary.

The final two lines of the input are the starting values to be used in the two trial runs. For the first solution initial values of .5 are assigned to each of the two paths *a* and *b* at the start of the iteration process; the second solution starts from the trial values .2 and .2.

If one were to run this problem with IPSOL, it would print out .602 and .410 as the final values for the solution on each of the two runs, values corresponding to line 6 in the hand solution in Table 2-1, with a minimum function value (sum of squared differences) of .0004469, again corresponding to the hand solution. The solution from initial values of .2, .2 would take a little longer, as it would take 6 cycles of MINIM to reach trial values of .6 and .4 (reached in 2 cycles from .5, .5); thereafter the solutions proceed identically.

Geographies of search

For the simple 2-variable case of Fig. 2.1 and Table 2-1 we can visualize the solution process as a search of a geographical terrain for its lowest point. Values of *a* and *b* represent spatial coordinates such as latitude and longitude, and values of the criterion Σd^2 represent altitudes above sea level. Fig. 2.2 is a pictorial representation of the situation. A set of starting trial values represents the coordinates of a starting point in the figure. The tests in steps a and b in each cycle represent tests of how the ground slopes each way from the present location, which governs the choice of a promising direction in which to move. In each instance we make the move that takes us downhill most rapidly. Eventually, we reach the low point in the valley, marked by the arrow, from which a step in any direction would lead upward. Then we quit and report our location as the solution.

Note that in simple geographies, such as that represented in this example, it doesn't matter what set of starting values we use--we would reach the same final low point regardless of where we start from--at worst it will take longer from some places than from others. Not all geographies, however, are

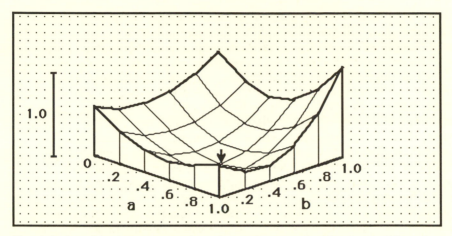

Fig. 2.2 Graphical representation of search space for Fig. 2.1 problem, for values 0 to 1 of both variables.

this simple. Fig. 2.3 shows a cross-section of a more treacherous terrain. A starting point at A on the left of the ridge will lead away from, not towards, the solution--the searcher will wind up against the boundary at B. From a starting point at C, on the right, one will see initial rapid improvement but will be trapped at an apparent solution at D, well short of the optimum at E. Or one might strike a level area, such as F, from which no direction of initial step leads to improvement. Other starting points, such as G and H, will, however, lead satisfactorily to E.

It is ordinarily prudent, particularly when just beginning to explore the landscape implicit in a particular path model, to try at least two or three widely dispersed starting points from which to seek a minimum. If all the solutions converge on the same point and it represents a reasonably good fit to the data,

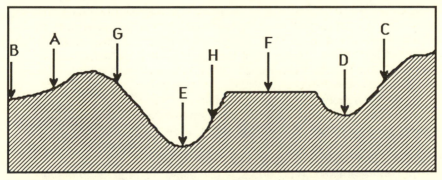

Fig. 2.3 Cross section of a less hospitable search terrain.

it is probably safe to conclude that it is the optimum solution. If some solutions wander off or stop short of the best achieved so far, it is well to suspect that one may be dealing with a less regular landscape and try additional sets of starting values until several converge on the same minimum solution.

The landscapes of Figures 2.2 and 2.3 are for problems in two unknowns. In the general case of *n* unknowns, the landscape would be an *n*-dimensional space with an *n*+1st dimension for the criterion. Although such spaces are not easily visualizable, they work essentially like the 3-dimensional one, with *n*-dimensional analogues of the valleys, ridges, and hollows of a 3-dimensional geography. The iterative procedure of Table 2-1 is easily extended to more dimensions (= more unknowns), although the amount of computation required escalates markedly as the number of unknowns to be solved for increases.

Many fine points of iterative minimization programs have been skipped over in this brief account. Some programs allow the user to place constraints on the trial values (and hence on the ultimate possible solutions), such as specifying that they always be positive, or that they lie between +1 and -1 or other defined limits. Programs differ in how they adjust their step sizes during their search, and in their ability to recover from untoward events (such as a trial value of zero appearing in the denominator of an expression). Some are extremely fast and efficient in certain kinds of friendly terrain but are not well adapted elsewhere. Others are robust, but painfully slow even on easy ground. Some programs allow the user a good deal of control over aspects of the search process and provide a good deal of information on how it proceeds. Others require a minimum of specification from the user and just print out a final answer.

Matrix Formulation of Path Models

Simple path diagrams are readily transformed into sets of simultaneous equations by the use of Wright's rules. We have seen in the preceding section how such sets of equations can be solved iteratively by computer programs. To use such a program, e.g., IPSOL, for this purpose, one must give it a subroutine containing the path equations, so that it can calculate the implied values and compare them with the observed values. With three observed values, as in our example, this is simple enough, but with 30 or 40 the preparation of a new subroutine for each problem can get rather tedious. Furthermore, in tracing paths in more complex diagrams to reduce them to sets of equations, it is easy to make errors--for example, to overlook some indirect path that connects point A and point B, or to include a path twice. Is there any way of mechanizing the construction of path equations, as well as their solution?

In fact, there are such procedures, which allow the expression of the equations of a path diagram as the product of several matrices. Not only does such an approach allow one to turn a path diagram into a set of path equations with less risk of error, but in fact one need not explicitly write down the path equations at all--one can carry out the calculation of implied correlations directly via operations on the matrices. This does not save effort at the level of actual computation, but it constitutes a major strategic simplification.

The particular procedure we use to illustrate this is one based on a formulation by McArdle and McDonald (1984); an equivalent although different matrix procedure is carried out within the computer program LISREL (of which more later), and still others have been proposed (e.g., Bentler & Weeks, 1980; McArdle, 1980; McDonald, 1978). It is assumed that the reader is familiar with elementary matrix operations; if your skills in this area are rusty or nonexistent, you may wish to consult Appendix A or an introductory textbook in matrix algebra before proceeding.

McArdle and McDonald define three matrices, **A**, **S**, and **F**:

A (for "asymmetric" relations) contains paths.

S (for "symmetric" relations) contains correlations (or covariances) and residual variances.

F (for "filter" matrix) selects out the observed variables from the total set of variables.

If there are a total of t variables (excluding residuals), m of which are measured, the dimensions of these matrices are: $\mathbf{A} = t \times t$; $\mathbf{S} = t \times t$; $\mathbf{F} = m \times t$.

The implied correlation (or covariance) matrix **C** among the measured variables is obtained by the matrix equation:

$$\mathbf{C} = \mathbf{F}\,(\mathbf{I} - \mathbf{A})^{-1}\,\mathbf{S}\,(\mathbf{I} - \mathbf{A})^{-1\,\prime}\,\mathbf{F}'.$$

I stands for the identity matrix, and $^{-1}$ and $'$ refer to the matrix operations of inversion and transposition, respectively.

This is not a very transparent equation. You may wish just to take it on faith, but if you want to get some sense of why it looks like it does, you can turn to Appendix C, where it is shown how this matrix equation can be derived from the structural equation representation of a path diagram. The fact that the equation does what it claims to do is shown by examples in the next two sections of this chapter.

An example with correlations

Figure 2.4 and Tables 2-3 and 2-4 provide an example of the use of the McArdle-McDonald matrix equation. The path diagram in Fig. 2.4 is that of Fig. 1.22, from the exercises of the preceding chapter.

Variables B, C, and D are assumed to be observed; variable A to be latent. All variables are assumed to be standardized--i.e., we are dealing with a correlation matrix. Expressions for the correlations and variances, based on path rules, are given to the right in the figure.

In Table 2-3, Matrix **A** contains the three straight arrows (paths) in the diagram, the two a s and the c. Each is placed at the intersection of the

$$r_{BC} = ba + c \qquad r_{BD} = ba$$
$$r_{CD} = a^2 + cba \qquad s_D^2 = a^2 + d^2$$
$$s_C^2 = c^2 + a^2 + 2abc + e^2$$

Fig. 2.4 A path diagram for the matrix example of Tables 2-3 and 2-4.

variable it originates from (top) and the variable it points to (side). For example, path c, which goes from B to C, is specified in row C of column B. It is helpful (though not algebraically necessary) to group together source variables and downstream variables--the source variables A and B are given first in the Table 2-3 matrices, with the downstream variables C and D last.

Curved arrows and variances are represented in matrix **S**. The top left-hand part contains the correlation matrix among the source variables, A and B. The diagonal in the lower right-hand part contains the residual variances of the downstream variables C and D, as given by the squares of the residuals paths e and d. (If there were any covariances among residuals, they would be shown by off-diagonal elements in this part of the matrix.)

Finally, matrix **F**, which selects out the observed variables from all the variables, has observed variables listed down the side and all variables along the top. It simply contains a 1 at the row and column corresponding to each observed variable--in this case, B, C, and D.

Table 2-4 demonstrates that multiplying out the matrix equation yields the path equations. First, **A** is subtracted from the identity matrix **I**, and the result inverted, yielding $(I-A)^{-1}$. You can verify that this *is* the required inverse by the

Table 2-3 Matrix formulation of a path diagram by the McArdle- McDonald procedure

A						S				
	A	B	C	D			A	B	C	D
A	0	0	0	0		A	1	b	0	0
B	0	0	0	0		B	b	1	0	0
C	a	c	0	0		C	0	0	e^2	0
D	a	0	0	0		D	0	0	0	d^2

F				
	A	B	C	D
B	0	1	0	0
C	0	0	1	0
D	0	0	0	1

Table 2-4 Solution of the McArdle-McDonald equation, for the matrices of Table 2-3

$(\mathbf{I}\text{-}\mathbf{A})^{-1}$

	A	B	C	D
A	1	0	0	0
B	0	1	0	0
C	a	c	1	0
D	a	0	0	1

$(\mathbf{I}\text{-}\mathbf{A})^{-1}\mathbf{S}$

	A	B	C	D
A	1	b	0	0
B	b	1	0	0
C	a+bc	ab+c	e^2	0
D	a	ab	0	d^2

$(\mathbf{I}\text{-}\mathbf{A})^{-1\,\prime}$

	A	B	C	D
A	1	0	a	a
B	0	1	c	0
C	0	0	1	0
D	0	0	0	1

$(\mathbf{I}\text{-}\mathbf{A})^{-1}\mathbf{S}\,(\mathbf{I}\text{-}\mathbf{A})^{-1\,\prime}$

	A	B	C	D
A	1	b	a+bc	a
B	b	1	ab+c	ab
C	a+bc	ab+c	$a^2+c^2+2abc+e^2$	a^2+abc
D	a	ab	a^2+abc	a^2+d^2

$\mathbf{F}(\mathbf{I}\text{-}\mathbf{A})^{-1}\mathbf{S}\,(\mathbf{I}\text{-}\mathbf{A})^{-1\,\prime}\,\mathbf{F}' = \mathbf{C}$

	B	C	D
B	1	ab+c	ab
C	ab+c	$a^2+c^2+2abc+e^2$	a^2+abc
D	ab	a^2+abc	a^2+d

matrix multiplication $(\mathbf{I}\text{-}\mathbf{A})^{-1}(\mathbf{I}\text{-}\mathbf{A}) = \mathbf{I}$. (If you want to learn a convenient way of obtaining this inverse, see Appendix C.) Pre and postmultiplying **S** by $(\mathbf{I}\text{-}\mathbf{A})^{-1}$ and its transpose is done in this and the next row of the table, and the result of pre and postmultiplying by **F** and **F**′ is shown in the last row. This final matrix contains expressions for the correlations among the observed variables. You will see that they agree with the results of applying Wright's rules to the path diagram.

The matrix to the right in the second row, $(\mathbf{I}\text{-}\mathbf{A})^{-1}\mathbf{S}\,(\mathbf{I}\text{-}\mathbf{A})^{-1\,\prime}$, contains the correlations among *all* the variables, both latent and observed. The final pre and postmultiplication by **F** merely selects out the lower right-hand portion of the matrix, the correlations among the observed variables. The upper left-hand portion contains the intercorrelations of the latent variables. The remaining upper right and lower left sections contain the correlations between the latent and observed variables. As you should verify, all these are consistent with those obtainable via path tracing on the diagram. Thus, with particular values of a, b, c, etc. inserted in the matrices, the matrix operations of the McArdle-McDonald equation result in exactly the same implied values for the intercorrelations as would putting these same values into expressions derived from the path diagram via Wright's rules.

An example with covariances

The only modification to the procedure that is needed in order to use it with a variance-covariance matrix is to insert variances instead of 1's in the upper diagonal of **S**. The equation will then yield an implied variance-covariance matrix of the observed variables, instead of a correlation matrix, with the path coefficients a and c in rawscore form.

The procedure is illustrated in Table 2-5. The example is the same as that in Table 2-4, except that variables A, B, C, and D are now assumed to be unstandardized. The table shows the **S** matrix (the **A** and **F** matrices are as in Table 2-3), and the final result. Notice that these expressions conform to the rawscore path rules, by the inclusion of one variance or covariance in each path, involving the variable or variables at its highest point. (The b's are, of course, covariances, and the a's and c's unstandardized path coefficients.) You may wish to check out some of this in detail to make sure you understand the process.

Table 2-5 Solution for covariance matrix, corresponding to Table 2-4

S

	A	B	C	D
A	s_A^2	b	0	0
B	b	s_B^2	0	0
C	0	0	e^2	0
D	0	0	0	d^2

$\mathbf{F(I-A)^{-1}S\,(I-A)^{-1}\,'\,F' = C}$

	B	C	D
B	s_B^2	$ab+c\,s_B^2$	ab
C	$ab+c\,s_B^2$	$a^2\,s_A^2+c^2\,s_B^2+2abc+e^2$	$a^2\,s_A^2+abc$
D	ab	$a^2\,s_A^2+abc$	$a^2\,s_A^2+d^2$

A Full-Fledged Model-Fitting Program--LISREL

Suppose one were to take a search program, such as IPSOL, and replace the path equations in the FUNCT subroutine with a matrix formulation, such as the McArdle-McDonald equation. By describing the matrices **A**, **S**, and **F** in the input to the program, one could avoid the necessity of writing fresh path equations and inserting them in the FUNCT subroutine for each new problem.

One might well dress up such a program with a few additional frills: For example, one could use a more sophisticated search program and speed up the matrix calculations in various ways. One could also offer some additional options in the way of criteria for evaluating goodness of fit. IPSOL minimizes the sum of squared differences between observed and implied correlations. This least squares criterion is one that is easily computed and widely used in statistics, but there are others, such as maximum likelihood, that might be considered and that could be provided as alternatives in subroutine FUNCT. (Some of the relative advantages and disadvantages of different goodness-of-fit criteria are discussed in a later section of this chapter.)

Several programs along these lines exist and can be used for solving path diagrams. One is COSAN, developed by R. P. McDonald of MacQuarie University in Australia. Another is EQS, developed by Peter Bentler of UCLA. A third program is MILS, developed by Ronald Schoenberg at the National Institutes of Health in Bethesda, Md. MILS is descended from an early version of the fourth program, LISREL, which is currently the most widely used of the group. LISREL, which stands for LInear Structural RELations, was devised by the Swedish psychometrician Karl Jöreskog, and has developed through a series of versions. The current version, LISREL VI (Jöreskog and Sörbom, 1984), is distributed as an optional procedure in the SPSS-X program package (SPSS, 1984).

LISREL is based on a more elaborate matrix formulation of path diagrams than the McArdle-McDonald equation, although one that works on similar principles and leads to the same end result. The LISREL formulation is more complicated because it subdivides the process, keeping in eight separate matrices various elements that are combined in the three McArdle-McDonald matrices. LISREL makes basic distinctions between independent and dependent variables in addition to latent and manifest variables. The paths connecting latent and manifest variables, and the residual paths for the latter, are collectively called the *measurement model.* There are two measurement submodels, one for the independent variables, and one for the dependent variables. The relationships among and between the corresponding two sets of latent variables are called the *structural model.* This is a formulation within which many problems of interest to social and behavioral scientists can easily and naturally be expressed, and which is flexible enough to handle quite a few others by various specializations and equivalences requiring greater or lesser amounts of ingenuity on the part of the user. We consider a number of examples in the next chapter.

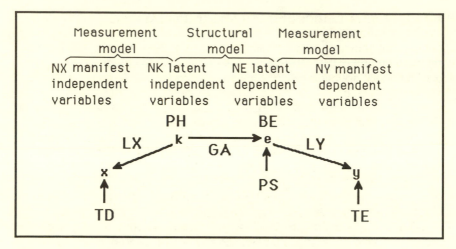

Fig. 2.5 Summary of matrix representation in LISREL.

Figure 2.5 summarizes the relationships in LISREL. The portions of the diagram to the left and the right constitute the two measurement submodels; the portion in the middle is the structural model. The latent independent variables K are source variables. There are NK of them, in a vector **k**. The latent dependent variables E are downstream variables; there are NE of them. There are NX observed independent variables X, and NY observed dependent variables Y. The X's and Y's are all considered to be downstream variables-- fallible observed indexes only imperfectly reflecting the true latent variables lying behind them whose relationships are our real concern. The vertical upward arrows in Fig. 2.5 represent the residual paths for the three sets of downstream variables--of course, there are none for the K's, which are source variables. The two-letter symbols next to the arrows in the diagram represent the eight matrices used by LISREL (in practice, many problems require only a subset of these). The matrices **LX** and **LY** represent the paths from latent to observed variables, and **TD** and **TE** the variance-covariance matrices of residuals (diagonal matrices, if the residuals are assumed to be uncorrelated). These four matrices constitute the measurement model. The dimensions of these matrices follow from the numbers of variables involved: **LX** is NX x NK, **LY** is NY x NE, and **TD** and **TE** are square matrices of order NX and NY, respectively.

In the structural model, **GA** contains paths from source to downstream latent variables. **PH** represents the variance-covariance matrix of the source variables, and **BE** the paths between downstream latent variables. **GA** is of dimension NE x NK, and **PH** and **BE** are of order NK and NE, respectively. **PS**, the variance-covariance matrix of the residuals of the downstream latent variables, is also a square matrix of order NE.

The LISREL input sets up the eight matrices with a combination of fixed values (usually 1's and 0's) and unknown parameters to be solved for. The multiplication of these matrices yields the implied values of the correlations (or covariances) that are used by the minimization part of the program. We need not go into the complete details of the matrix operations, but to take a simple case, the implied variances and covariances among the independent variables can be obtained by:

LX PH LX′ + TD.

An example is given in Table 2-6 based on the path diagram of Fig. 2.6. The

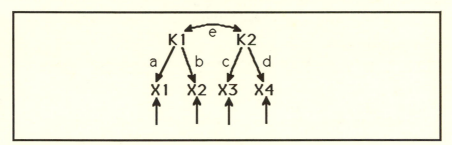

Fig. 2.6 Path diagram for example of Table 2-6.

Table 2-6 Example showing equivalence of path expressions and the matrix expression **LX PH LX′ + TD**

				Path expressions for correlations			
					X1	X2	X3	X4
PH				X1	1	ab	aec	aed
		K1	K2	X2	ab	1	bec	bed
	K1	1	e	X3	aec	bec	1	cd
	K2	e	1	X4	aed	bed	cd	1
LX				**TD**			
		K1	K2			X1	X2	X3	X4
	X1	a	0		X1	$1-a^2$	0	0	0
	X2	b	0		X2	0	$1-b^2$	0	0
	X3	0	c		X3	0	0	$1-c^2$	0
	X4	0	d		X4	0	0	0	$1-d^2$
LX PH				**LX PH LX′**			
		a	ae		a^2	ab	aec	aed
		b	be		ab	b^2	bec	bed
		ce	c		aec	bec	c^2	cd
		de	d		aed	bed	cd	d^2

path expressions derived from the diagram by Wright's rules are shown at the top right of Table 2-6; the LISREL matrices **PH**, **LX**, and **TD** are given to the left and below; and the calculation **LX PH LX´** is shown. The result plus **TD** clearly will equal the correlation matrix at the top.

Table 2-7(a) shows one way of defining input to LISREL for the problem of Fig. 2.6 and Table 2-6. The first line of the input is a title. The second gives data specifications: number of input variables = 4; number of observations (cases) = 100; the matrix to be analyzed is a correlation matrix (CM would mean covariance matrix). The third line specifies that a correlation matrix is to be read, in free-field format (items separated by any number of spaces). If the matrix were to be in other than lower triangular form, or in a fixed format, this would be specified as well. The correlations follow in lines 4 through 7.

Table 2-7 Two examples of LISREL input

(a)

```
INPUT FOR FIG 2.6 PROBLEM
DATA NI=4  NO=100  MA=KM
KM
   1.00
    .50  1.00
    .10   .10  1.00
    .20   .30   .20  1.00
MODEL NX=4  NK=2  PH=ST
FREE  LX 1 1 LX 2 1 LX 3 2 LX 4 2
START  .5  ALL
OUTPUT NS
```

(b)

LX	LY	PH	GA	PS	TD		TE	
a	1	1	d	v	w	0	y	0
b	c				0	x	0	z

```
INPUT FOR FIG 2.7 PROBLEM
 [lines 2-7 same as for (a) above]
SELECT
3 4 1 2
MODEL NX=2  NK=1  NY=2  NE=1  PH=ST  LX=FR  LY=FR
FIX LY 1 1
START 1.0 LY 1 1
OUTPUT SS  UL
```

Note: Details of input vary slightly between different LISREL versions, these are for LISREL VI.

The lines from MODEL to START define the matrices and starting values of the model. The general philosophy of LISREL is that things are assumed to be in some typical form by default unless otherwise specified. The MODEL line says that there are 4 X variables and 2 K variables. The "ST" specifies that the **PH** matrix is to be a correlation matrix (symmetrical, with 1's fixed in the diagonal and free values elsewhere--the *e*'s of Table 2-6). Because nothing is said about **LX** and **TD**, these matrices are assumed to be in their default forms--respectively, a rectangular matrix with all values fixed to zero, and a diagonal matrix with all diagonal values free. The line beginning FREE specifies exceptions: **LX** locations 1 and 2 of column 1 and 3 and 4 of column 2 are to be free values to be solved for (they correspond to *a, b* and *c, d* of Table 2-6). The line beginning START sets an initial value of .5 into all free values. The NS in the OUTPUT line tells LISREL to use the starting values we have provided, rather than calculate its own. Because OUTPUT carries no additional specifications the standard output will be produced and the standard fitting criterion will be used (for LISREL this is maximum likelihood).

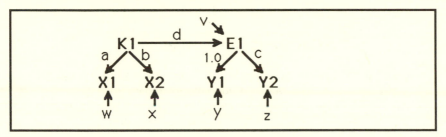

Fig. 2.7 A different model for the data of Fig. 2.6.

Figure 2.7 shows a different model that might be fit to the same data. In this model, we assume that there are two X variables, the same X1 and X2 as in Fig. 2.6, but now X3 and X4 are considered to be dependent variables, Y1 and Y2, and there is a causal path, labeled *d*, rather than a simple correlation, between the two latent variables. Also, because this makes K2 into E1, a downstream variable, it acquires a residual arrow.

Table 2-7 (b) shows the modifications in input required. There are now seven LISREL matrices, as shown. (**BE** is the unused eighth--there are no arrows between two downstream latent variables.)

Aside from the title, there are no changes in the first seven lines of the input, through the correlation matrix. The next two lines reorder the input: LISREL requires that dependent variables (the third and the fourth in the correlation matrix as read) precede independent variables (1 and 2). The MODEL line indicates two X and two Y variables, and one K and one E. **LX** and **LY** are started as FREE matrices, and then, in the next two lines, the first element of **LY** is fixed and set to 1. This illustrates an important LISREL feature: Downstream latent variables cannot be specified as standardized.

Instead, they must be given an arbitrary scale in the initial solution by assigning a fixed value to a path leading to an observed variable--in this case, Y1. Afterwards, however, LISREL can standardize all the latent variables and will if requested to do so by an SS in the OUTPUT line, for "standardized solution." (*Source* latent variables may be treated either way.) The UL in the OUTPUT line requests that fitting be done using an ordinary ("unweighted") least squares criterion. This time LISREL is permitted to calculate its own start values.

Many input variations are possible in LISREL. Matrices specifying the model can be read in explicitly, if desired; or one can specify matrices to be all zeroes, diagonal, symmetrical, and so on. Lines of the form EQUAL LX 1 1 LX 2 2 can be used to specify that LX 2 2 is constrained to be equal to LX 1 1. Various amounts of detail can be requested in the output.

A serious user of LISREL will need to consult the relevant manuals for details; these illustrations are merely intended to convey something of the flavor of LISREL's style, and to provide models for working simple problems. A number of examples of the use of LISREL in actual research are found in the next two chapters.

Goodness-of-Fit Criteria

Earlier in this chapter, it was mentioned that different criteria may be used to assess how well the correlation or covariance matrix implied by a particular set of trial values fits the observed data. Three such criteria are described in this section, ordinary least squares, generalized least squares, and maximum likelihood.

Why three criteria? The presence of more than one places the user, to some extent, in the situation described in the proverb: A man with one watch always knows what time it is; a man with two watches never does. The answer is, that the different criteria have different advantages and disadvantages, as we see shortly. We may formulate the three criteria in matrix terms as follows, where OLS, GLS, and ML stand for ordinary least squares, generalized least squares, and maximum likelihood, **S** and **C** are the m x m observed and implied covariance or correlation matrices, *tr* refers to the *trace*, the sum of the diagonal elements of a matrix, *ln* is a natural logarithm, and vertical lines indicate a determinant. The three criteria may be given as:

$$OLS = tr(\mathbf{S} - \mathbf{C})^2$$
$$GLS = 1/2 \ tr \ [(\mathbf{S} - \mathbf{C}) \ \mathbf{S}^{-1}]^2$$
$$ML = ln \ |\mathbf{C}| - ln \ |\mathbf{S}| + tr \ \mathbf{S}\mathbf{C}^{-1} - m$$

Although not obvious without a little study, the expression for the first criterion is just a way of describing the sum of the squares of the differences of the corresponding elements of the **S** and **C** matrices. (The procedure works because **S - C** is a symmetric matrix.)

Neglecting the scaling factor of 1/2, the expression for generalized least squares is like that for ordinary least squares, except that the matrix of differences between **S** and **C** is multiplied by the inverse of **S** before squaring. (In an alternative version of GLS, $\mathbf{C}^{-1}$ is used instead of $\mathbf{S}^{-1}$ as the multiplier.)

The details of the maximum likelihood function go a bit beyond our scope here, except that the reader should note that it consists of two parts: the difference between the natural logarithms of the determinants of the **C** and **S** matrices, and the difference between tr $\mathbf{SC}^{-1}$ and m, the order of the **S** and **C** matrices. It should be evident that if **S** and **C** are identical, they will have the same determinants, and the first part of the expression for ML will be zero. Also, if **S** and **C** are identical, $\mathbf{SC}^{-1}$ will be an identity matrix, the sum of its diagonal elements will be m, and the second part of the expression for ML will also be zero. It is not obvious, although it is true, that whenever **S** and **C** differ the sum of the two components will always be positive; that is, the ML function takes on its lowest possible value, zero, when **S** and **C** are identical.

It should be evident that the OLS criterion, because it is a sum of squared differences, also has a minimum value of zero that occurs when **S** and **C** are identical. It should be evident also that the GLS criterion will be zero in this case, because the expression within brackets will then be a null matrix, although it is less obvious that it must be positive otherwise.

Table 2-8 illustrates the calculation of the three criteria for two **C** matrices departing slightly from **S** in opposite directions. The upper part of the table illustrates the steps for calculating ML; the lower part shows OLS and GLS. Note that all of the goodness-of-fit criteria are small, reflecting the closeness of **C** to **S**, and that they are positive for either direction of departure from **S**. It happens that the ML and GLS criteria are the same in this case to two significant digits; as we see later, this is by no means always the case, although ML and GLS are in general scaled comparably (OLS is not).

Another message of Table 2-8 is that considerable numerical accuracy is required for calculations such as these--one more reason for letting computers do them. In this problem, a difference between **C** and **S** matrices in the second decimal place requires going to the sixth decimal place in the GLS and ML criteria in order to detect its effect. With only, say, 4 or 5-place accuracy in obtaining the logarithms and inverses, quite misleading results would have been obtained.

Goodness-of-fit criteria serve two purposes in iterative model fitting. First, they guide the search for a best fitting solution. Second, they evaluate the solution when it is obtained. The three criteria being considered have somewhat different relative merits for these two tasks.

For the first purpose, guiding a search, a criterion should ideally be cheap to compute, because the function is evaluated repeatedly at each step of a trial-and-error search. Furthermore, the criterion should be a dependable guide to relative distances in the search space, especially at points distant from a perfect fit. For the second purpose, evaluating a best fit solution, the statistical properties of the criterion are a very important consideration, computational

Table 2-8 Sample computations of ML, OLS, and GLS criteria for the departure of covariance matrices C_1 and C_2 from S

	S		S^{-1}	
	2.00	1.00	.5714286	-.1428571
	1.00	4.00	-.1428571	.2857143
$|S|$	7.00			
ln $|S|$	1.9459101			

	C_1		C_2	
	2.00	1.00	2.00	1.00
	1.00	4.01	1.00	3.99
$|C|$	7.02		6.98	
ln $|C|$	1.9487632		1.9430489	
ln $|C|$ - ln $|S|$	.0028531		-.0028612	
C^{-1}	.5712251	-.1424501	.5716332	-.1432665
	-.1424501	.2849003	-.1432665	.2865330
SC^{-1}	1.0000000	.0000000	1.0000000	.0000000
	.0014245	.9971510	-.0014327	1.0028653
tr SC^{-1}	1.9971510		2.0028653	
tr SC^{-1}- m	-.0028490		.0028653	
ML	.0000041		.0000041	
$S - C$	.00	.00	.00	.00
	.00	-.01	.00	.01
OLS	.0001000		.0001000	
$(S-C)\,S^{-1}$	.0000000	.0000000	.0000000	.0000000
	.0014286	-.0028571	-.0014286	.0028571
$[(S-C)\,S^{-1}]^2$	.0000000	.0000000	.0000000	.0000000
	.0000041	.0000082	-.0000041	.0000082
GLS	.0000041		.0000041	

cost is a minor issue, and the behavior of the function in remote regions of the search space is not in question.

In computational cost, ordinary least squares is by all odds the cheapest, GLS is second, and ML requires the most computation. The difference between the latter two is primarily because ML uses C^{-1} and GLS uses S^{-1}. S, the observed correlation matrix, remains the same throughout the computations, so it is only necessary to invert it once at the beginning; whereas C changes with every change in the trial values of the unknowns, so it is necessary to recompute C^{-1} repeatedly. (The alternative version of GLS, which uses C^{-1} instead of S^{-1} as a multiplier, incurs this same computational cost.)

So far as statistical merits go, ML and GLS share an important advantage that OLS does not. For ML and GLS, the value of the criterion at the point of best fit, multiplied by N-1, where N is the number of subjects, yields a quantity that, under appropriate conditions, approximately follows the χ^2 distribution. This permits statistical tests of goodness of fit in a manner described in the next section.

If the original data follow a multivariate normal distribution, the ML criterion is appropriate, and will tend to have slightly better statistical properties than GLS, but GLS yields an approximate chi square test under somewhat less restrictive assumptions than multivariate normality (Browne, 1977). (The version of GLS that uses C^{-1} as a weight more closely approaches the statistical properties of ML--as well as its computational cost.)

In general, these statistical claims are based on the assumption of large samples. It is hard to say how large "large" is, because, as usual, things are not all-or-nothing--approximations get gradually rougher as sample size decreases; there is no single value marking a sharp boundary between smooth sailing and disaster. Also, other things being equal, simple-minded models with few variables can often get by with smaller sample sizes than are required by complex many-variable models. As a rough rule of thumb, one would probably do well to be modest in one's statistical claims even with fairly simple models if N is less than 100, and if more complex model fitting is contemplated, an N four or five times that size would not be an extravagance.

Finally, the criteria differ in their ability to provide dependable distance measures, especially at points remote from the point of perfect fit. On the whole, experience suggests that ML is most vulnerable to such difficulties, and OLS least.

Figure 2.8 and Table 2-9 provide an example of a case where ML gives an anomalous solution. The data are from Dwyer (1983, p. 258), and they represent the variance-covariance matrix for three versions of an item on a scale measuring authoritarian attitudes. The question Dwyer asked is whether the items satisfy a particular psychometric condition known as "tau-equivalence," which implies that they measure a single common factor for which they have equal weights, but possibly different residual variances, as shown in the path diagram of Fig. 2.8. It is thus a problem with four unknowns,

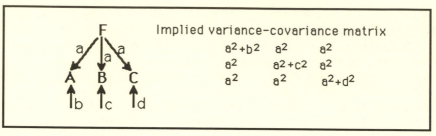

Fig. 2.8 Model of single common factor with equal loadings, plus different specifics ("tau-equivalent" tests).

a, b, c, and *d.* Such a model implies that the off-diagonal elements in **C** must all be equal, and so *a* should be assigned a value to give a reasonable compromise fit to the three covariances. The unknowns *b, c,* and *d* can then be given values to insure a perfect fit to the three observed values in the diagonal.

This is just what an iterative search program using an OLS criterion does, as shown in the left column of Table 2-9. A value of $\sqrt{5.58}$ is found for *a,* and

Table 2-9 OLS and ML solutions for Fig. 2.8

	S					
	6.13	6.12	4.78			
	6.12	8.29	5.85			
	4.78	5.85	7.35			
	C$_{OLS}$			**C**$_{ML}$		
	6.13	5.58	5.58	6.46	5.66	5.66
	5.58	8.29	5.58	5.66	7.11	5.66
	5.58	5.58	7.35	5.66	5.66	8.46
S - C	.00	.54	-.80	-.33	.46	-.88
	.54	.00	.27	.46	1.18	.19
	-.80	.27	.00	-.88	.19	-1.11
|S|		41.2902			41.2902	
|C|		43.1529			45.4716	
ln |**C**| - ln |**S**|		.0441			.0965	
tr **SC**$^{-1}$		3.2751			3.0028	
tr **SC**$^{-1}$-m		.2751			.0028	
ML		.3192			.0993	
OLS		2.0090			4.7776	

values of $\sqrt{.55}$, $\sqrt{2.71}$, and $\sqrt{1.77}$ for *b*, *c*, and *d*, respectively, yielding the implied matrix $\mathbf{C}_{OLS}$. Dwyer used a ML criterion (with the program LISREL) and obtained a solution giving the implied matrix on the right in Table 2-9, labeled $\mathbf{C}_{ML}$. Notice that this matrix has equal off-diagonal values, as it must, but that the diagonal values are not at all good fits to the variances in **S**, as shown by the matrix **S-C**. The values of the ML criterion for the two **C** matrices are given in the bottom part of the table. It is clear that the ML goodness-of-fit criterion for $\mathbf{C}_{ML}$ is very substantially smaller than that for the solution which the eye and OLS would judge to be superior. $\mathbf{C}_{ML}$ happens to be a matrix for which tr $\mathbf{SC}^{-1}$ is very nearly equal to 3.00, and this is more than enough to offset the advantage $\mathbf{C}_{OLS}$ has over $\mathbf{C}_{ML}$ with respect to the difference $\ln|\mathbf{C}| - \ln|\mathbf{S}|$.

Also shown in Table 2-9 are OLS criteria for the fit of $\mathbf{C}_{OLS}$ and $\mathbf{C}_{ML}$ to **S**. As one might expect, the OLS criterion is considerably smaller for the solution $\mathbf{C}_{OLS}$ than for $\mathbf{C}_{ML}$.

Table 2-10 gives a further example to illustrate that the three criteria do not always agree on the extent to which one covariance matrix resembles another, and that ML, in particular, is sometimes a very erratic judge of distance when distances are not small. In the table, two different **C** matrices are evaluated relative to the **S** matrix shown at the top. Recall that the scale of OLS is not necessarily commensurate with those of GLS and ML, so that it is the relative rather than absolute differences that should be compared.

The two least squares criteria agree that matrix $\mathbf{C}_2$ is much less like **S** than matrix $\mathbf{C}_1$ is, but ML actually asserts the opposite: that $\mathbf{C}_1$ is more different from **S** than is $\mathbf{C}_2$.

This is not to say that ML will not give accurate assessments of fit when the fit is good, that is, when **C** and **S** are close to each other. In Table 2-8 the OLS and GLS criteria agreed very well for **C**'s differing only very slightly from

Table 2-10 How three goodness-of-fit criteria evaluate the resemblance of two Cs to S

Matrix			OLS	GLS	ML
S	2	1			
	1	4			
$\mathbf{C}_1$	1	2	4.0	.449	6.054
	2	5			
$\mathbf{C}_2$	11	7	154.0	10.755	4.513
	7	5			

this same **S**. But in the early stages of a search when **C** is still remote from **S**, or for problems like that of Table 2-9 where the best fit is not a very good fit, ML's eccentricities can give trouble. After all, if a fitting program were to propose C_1 as an alternative to C_2 (Table 2-10), OLS and GLS would accept it as a dramatic improvement, but ML would reject it as tending to move away from **S**!

None of this is meant to imply that searches using the ML criterion are bound to run into difficulties--in fact, studies reviewed in the next section suggest that ML in practice often works quite well. I do, however, want to emphasize that uncritical acceptance of any solution a computer program happens to produce can be hazardous to one's scientific health. If in doubt, one should try solutions from several starting points with two or three different criteria--if all converge on similar answers, one can then use the ML solution for its favorable statistical properties, or, if one is dubious about multivariate normality, GLS.

Monte Carlo studies using the maximum likelihood criterion

There have been a couple of recent studies in which fairly extensive explorations have been made with the program LISREL and a maximum likelihood criterion, studies based on repeated random sampling from artificial populations with known characteristics (Anderson & Gerbing, 1984; Boomsma, 1982, 1985; Gerbing & Anderson, 1985). These studies manipulated model characteristics and sample sizes and studied the effects on accuracy of estimation and the frequency of improper or nonconvergent solutions.

These studies do not address all possible issues, of course. Anderson and Gerbing worked solely with confirmatory factor analysis models, and Boomsma largely did, so the results apply most directly to models of this kind. Both studies sampled from multivariate normal populations, so questions of the robustness of maximum likelihood to departures from multivariate normality are not addressed. For the most part, both studies used optimum starting values for the iteration, namely, the true population values; thus, the behavior of the maximum likelihood criterion in regions distant from the solution is not at issue. (In one part of her study, Boomsma did compare alternative starting points.)

Within these limitations, a variety of models and sample sizes were used in the two studies combined. The number of latent variables (factors) ranged from 2 to 4, and the correlations between them were .0, .3, or .5. The number of observed indicators per latent variable ranged from 2 to 4, and the sizes of nonzero factor pattern coefficients from .4 to .9, in various combinations. Sample sizes of 25, 50, 75, 100, 150, 200, 300, and 400 were employed.

The main tendencies of the results can be briefly summarized, although there were some complexities of detail for which the reader may wish to consult the original articles.

First, *convergence failures*. These occurred quite frequently with small samples and few indicators per factor. In fact, with samples of less than 100 cases and only two indicators per factor, such failures occurred on almost half the trials under some conditions (moderate loadings and low interfactor correlations). With three or more indicators per factor and 150 or more cases, failures of convergence rarely occurred.

Second, *improper solutions* (negative estimates of residual variance-- so-called "Heywood cases"). Again, with samples of less than 100 and only two indicators per factor, these cases were very common. With three or more indicators per factor and sample sizes of 200 or more, they were pretty much eliminated.

Third, *accuracy.* With smaller samples, naturally, estimates of the population values were less precise--that is, there was more sample- to-sample variation in repeated sampling under a given condition. However, with some exceptions for the very smallest sample sizes (25 and 50 cases), the standard error estimates provided by the LISREL program appeared to be dependable-- that is, a 95% confidence interval included the population value somewhere near 95% of the time.

Finally, *starting points.* As mentioned, Boomsma in part of her study investigated the effect of using alternative starting values. This aspect of the study was confined to otherwise favorable conditions--samples of 100 or more cases with three or more indicators per factor--and the departures from the ideal starting values were not *very* drastic. Under these circumstances, the solutions usually converged, and when they did it was nearly always to essentially identical final values; differences were mostly in the third decimal place or beyond.

In summary, then, these studies suggest that for the class of problems investigated--confirmatory factor analyses with two to four factors--one should plan on sample sizes of at least 100 cases (200 is better) and use at least three indicators per factor. Using a combination of two indicators per factor and 50 cases or less is just inviting trouble. In between there are tradeoffs. If you are stuck with a small sample, use more indicators per factor if you can and plan towards replication in another sample.

The results reported were obtained using LISREL and multivariate normal distributions, but presumably they would serve as guidelines for other similar programs, and distributions more or less approximating a multivariate normal form.

We encounter here, of course, the usual dilemma confronting the individual who would like to be both statistical purist and practical researcher. Few, if any, users of chi square tests or standard error estimates with maximum likelihood estimation are in a position to fully justify the probability values they report. Nearly always, the strongest appropriate claim would be something like: "To the extent that the underlying assumptions hold, we can conclude that. . . . " Or, perhaps more frankly: "The statistical tests and probability values in this paper are reported in a mainly descriptive spirit, to help orient the reader among the various models we present."

Some such statement should be taken as appended to every substantive study described in this book. I have not ordinarily made it explicitly. To do it only on occasion would be invidious. To do it every time would be an unkindness to the reader.

Hierarchical χ^2 Tests

As noted earlier, for GLS or ML, but not for OLS, one can multiply the criterion at the point of best fit by N -1 to obtain an approximate χ^2 in large samples. This can be used to test the fit of the implied **C** to **S**. The degrees of freedom for the comparison are the number of independent values in **S** less the number of unknowns used in generating **C**.

For example, in the problem of tau-equivalence discussed in the preceding section (Fig. 2.8), there were $m (m +1)/2 = 6$ independent values in **S** (the three variances in the diagonal and the three covariances on one side of it). There were four unknowns being estimated, *a*, *b*, *c,* and *d.* So there are two degrees of freedom for a χ^2 test. The minimum value of the ML criterion is .0993 (Table 2-9). As it happens, the data were gathered from 109 subjects, so $\chi^2 =108$ x .0993 = 10.7. From a χ^2 table (see Appendix G), the χ^2 with 2 df required to reject the null hypothesis at the .05 level is 5.99. The obtained χ^2 of 10.7 is larger than this, so we would reject the null hypothesis and conclude that the model of tau-equivalence did not fit these data; that is, that the difference between **C** and **S** is too great to readily be attributed to sampling error. Notice that the χ^2 test is used to conclude that a particular model *does not* fit the data. Suppose that χ^2 in the preceding example had been less than 5.99, what could we then have concluded? We could not conclude that the model is correct, but merely that our test had not shown it to be incorrect. How impressive this statement is depends very much on how powerful a test we have applied. By using a sufficiently small sample, for instance, we could fail to reject models that are grossly discrepant from the data. On the other hand, if our sample is extremely large, a failure to reject the model would imply a near-exact fit between **C** and **S**. Indeed, with very large samples we run into the opposite embarrassment, in that we may obtain highly significant χ^2s and hence reject models in cases where the discrepancies between model and data, although presumably not due to chance, are not large enough to be of any practical concern. It is prudent always to examine the residuals **S** - **C**, in addition to carrying out a χ^2 test, before coming to a conclusion about the fit of a model.

It is also prudent to look at alternative models. The fact that one model fits the data reasonably well does not mean that there could not be other, different models that fit better. At best, a given model represents a tentative explanation of the data. The confidence with which one accepts such an explanation

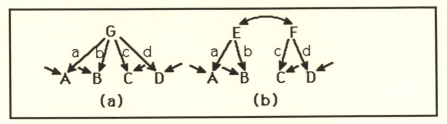

Fig. 2.9 Path models for the χ^2 comparisons of Table 2-11.

depends, in part, on whether other, rival explanations have been tested and found wanting.

Figure 2.9 and Table 2-11 provide an example of testing two models for fit to an observed set of intercorrelations, given as **S** at the top of the table. Model (a) is a Spearmanian model with a single general factor G. Model (b) has two correlated common factors, E and F. Fits to the data, using an iterative solution with a maximum likelihood criterion, are shown for both models. If we assume that the correlations in **S** are based on 120 subjects, what conclusions should we draw?

As the individual χ^2s for the two models indicate, we can reject neither. The correlation matrix **S** could represent the kind of chance fluctuation to be expected in random samples of 120 cases drawn from populations where the true underlying situation was that described by either model (a) or model (b).

Suppose that the correlations had instead been based on 240 subjects. Now what conclusions would be drawn?

Table 2-11 Comparing two models with χ^2

S	1.00	.30	.20	.10
	.30	1.00	.20	.20
	.20	.20	1.00	.30
	.10	.20	.30	1.00

	model		
	(a)	(b)	(diff)
χ^2, N = 120	4.64	.75	3.89
χ^2, N = 240	9.31	1.51	7.80
df	2	1	1
$\chi^2_{.05}$	5.99	3.84	3.84

In this case, we could reject model (a) because its χ^2 exceeds the 5.99 required to reject the null hypothesis at the .05 level with 2 *df*.   Model (b), however, remains a plausible fit to the data.

Does this mean that we can conclude that model (b) fits significantly better than model (a)? Not as such--the fact that one result is significant and another is nonsignificant is *not* the same as demonstrating that there is a significant difference between the two, although, regrettably, one sees this error made fairly often. (If you have any lingering doubts about this, consider the case where one result is just a hairsbreadth below the .05 level and the other just a hairsbreadth above--one result is nominally significant and the other not, but the difference between the two is of a kind that could very easily have arisen by chance.)

There is, however, a direct comparison that can be made in the case of Table 2-11 because the two models stand in a *nested*, or hierarchical, relationship. That is, the model with the smaller number of free variables can be obtained from the model with the larger number of free variables by fixing one or more of the latter. In this case, model (a) can be obtained from model (b) by fixing the value of the the interfactor correlation *e* at 1.00--if E and F are standardized and perfectly correlated, they can be replaced by a single factor G.

Two such nested models can be compared very simply by a χ^2 test: The χ^2 for this test is just the difference between the separate χ^2s of the two models, and the df is just the difference between their dfs (which is equivalent to the number of parameters fixed in going from the one to the other). In the example of Table 2-11, the difference between the two models turns out in fact to be statistically significant, as shown in the rightmost column at the bottom of the table. Interestingly, this is true for either sample size. In this instance, with N = 120 either model represents an acceptable explanation of the data, but model (b) provides a significantly better one than does model (a).

Figure 2.10 further illustrates the notion of nested models. Models 1,2, 3, and 4 represent such a hierarchical series because 2 can be obtained from 1 by setting path *c* to the fixed value of zero, 3 from 2 by similarly fixing *d*, and 4 from 3 by fixing *a* and *e* to zero. Obviously, in such a series any lower model can be obtained from any higher one by fixing paths--e.g., model 4 can be obtained from model 1 by setting paths *a*, *c*, *d*, and *e* to zero. Thus tests based on differences in χ^2 can be used to compare the fit of any two models in such a nested series. In the last described case such a test would have four degrees of freedom, corresponding to the four paths fixed in going from model 1 to model 4.

However, models 5, 6, and 7 in Fig. 2.10, while hierarchically related to model 1 and each other, are not in the same series as 2, 3, and 4. Thus, model 6 could not be compared with model 3 by taking the difference in their respective χ^2s. Although model 6 has fewer paths than model 3, they are not included within those of model 3--model 6 has path *c* as an unknown to be

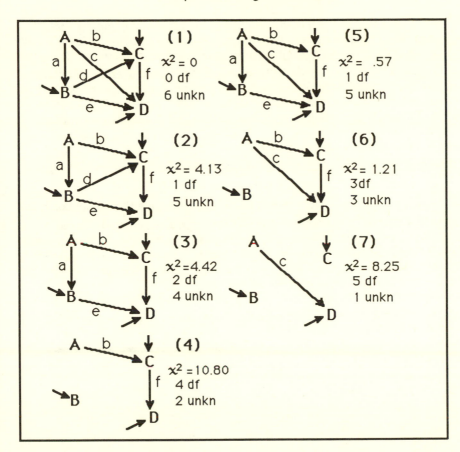

Fig. 2.10 Hierarchical series of path models (χ^2s hypothetical).

solved for, whereas model 3 does not.

Assuming that the four variables A, B, C, and D are all measured, model 1 is a case with $n(n-1)/2 = 6$ observed correlations and 6 unknowns to be solved for. A perfect fit will in general be achievable, χ^2 will be 0, and there will be 0 df. Obviously, such a model can never be rejected, but then, because it can be guaranteed to fit perfectly, its fit provides no special indication of its merit. The other models in Fig. 2.10 do have degrees of freedom and hence can potentially be rejected. Notice that the direct χ^2 tests of these models can be considered as special cases of the χ^2 test of differences between nested models because they are equivalent to the test of differences between these models and model 1.

Table 2-12 gives some examples of nested χ^2 tests based on the models of Fig. 2.10. The test in the first line of the table, comparing models 2 and 1, can be considered to be a test of the significance of path *c*. Does constraining path *c* to be zero significantly worsen the fit to the data? The answer, based on $\chi^2 = 4.13$ with 1 df, is yes. Path *c* makes a difference; the model fits significantly better with it included. Another test of the significance of a single path is provided in line 6 of the table, model 5 versus model 1. Here it is a test of the path *d*. In this case, the data do not demonstrate that path *d* makes a significant contribution: $\chi^2 = .57$ with 1 df, not significant. A comparison of model 3 with model 1 (line 2) is an interesting case. Model 2, remember, did differ significantly from model 1. But model 3, with one less unknown, cannot be judged significantly worse than 1 ($\chi^2 = 4.42$, 2df, NS). This mildly paradoxical situation arises occasionally in such χ^2 comparisons. It occurs because the increase in χ^2 in going from model 2 to model 3 is more than offset by the increase in degrees of freedom. Thus whereas the additional restriction on model 3 caused by setting path *d* to zero makes it fit worse than model 2 in an absolute sense, relative to the degrees of freedom one is less confident that the difference from model 1 is real.

Table 2-12 Some χ^2 tests for hierarchical model comparisons based on Fig. 2.10

	Model comparison	χ^2 1st	2nd	df 1st	2nd	χ^2_{diff}	df_{diff}	p
1.	2 vs 1	4.13	0	1	0	4.13	1	<.05
2.	3 vs 1	4.42	0	2	0	4.42	2	NS
3.	3 vs 2	4.42	4.13	2	1	.29	1	NS
4.	4 vs 3	10.80	4.42	4	2	6.38	2	<.05
5.	4 vs 1	10.80	0	4	0	10.80	4	<.05
6.	5 vs 1	.57	0	1	0	.57	1	NS
7.	6 vs 1	1.21	0	3	0	1.21	3	NS
8.	7 vs 6	8.25	1.21	5	3	7.04	2	<.05

Eliminating paths *a* and *e* in addition to *d* and *c* (model 4) is clearly going too far, whether the comparison is made with the full model (line 5), or with a model with *c* and *d* removed (line 4). However, if *c* is present *a* and *e* can be dispensed with--or at any rate, they cannot be demonstrated to be essential (line 7). Nevertheless, as line 8 indicates, *c* alone cannot do the job--deleting *b* and *f* leads to a significant worsening of fit ($\chi^2 = 7.04$, 2 df, p < .05).

Figure 2.10 also illustrates that the fact that a given model cannot be rejected does not mean that one should conclude that it represents the truth. Consider model 3. One cannot reject it as a plausible explanation of the data (χ^2 = 4.42, 2 df, NS--line 2). But this does not mean that other models might not do at least as well. Indeed, we have one in model 6, that with fewer parameters actually achieves a smaller χ^2. To be sure, we cannot carry out a direct statistical test of the relative goodness of fit of models 3 and 6 because they are not nested, but one would hardly wish to cheer very loudly about model 3 if one were aware that model 6 was lurking in the wings. The moral is that it pays to do some exploring of alternative models before going too far out on a limb on the basis of a significance test of any one. Otherwise, one risks the embarrassment of an unsuspected model 6 turning up.

Criteria for Exploratory Model Fitting

Although the χ^2 tests described in the preceding section are valuable and useful, they have some limitations when employed as a guide to exploratory model fitting. For one thing, the probability values become suspect when multiple model comparisons are being made--if enough models are tested, some failures to reject may occur purely by chance. For another thing, the role of sample size presents problems, as noted: With too-small samples, a poor fit may fail to be statistically rejected, and with too-large samples, a relatively good fit may still result in a significant χ^2. Several criteria have been proposed to address these problems. We begin by mentioning four based on χ^2: Jöreskog's χ^2/df, Bentler and Bonett's normed fit index, James, Mulaik, and Brett's parsimonious fit index, and Akaike's information criterion.

Jöreskog's χ^2/df

Jöreskog (Jöreskog & Sörbom, 1979, p. 39) suggests the use of the ratio of χ^2 to its degrees of freedom as a guide to model comparisons during exploratory model fitting. If the ratio is less than 1.0, it may be an indication that the fit of the model is "too good, " due to capitalization on chance during the model-fitting procedure. On the other hand, if the ratio of χ^2 to df is large, it suggests that improvement should be possible, and that further models with more or different paths should be considered. Jöreskog does not provide any specific value for what "large" is--it would, as he emphasizes, depend on the circumstances--but in several of the examples he gives, the dividing line seems to be at a χ^2/df ratio of about 2.

The first column in the right part of Table 2-13 gives χ^2/df values for the models of Fig. 2.10. Models 2, 3, 4, and the null model (described later) have relatively high ratios of χ^2 to degrees of freedom, suggesting that one could

67

Table 2-13 Comparison of four criteria for exploratory model-fitting, for models of Fig. 2.10

model	χ^2	df	unknowns	χ^2/df	nfi	pfi	AIC
				four criteria			
1	0.00	0	6		1.00	.00	-6.00
2	4.13	1	5	4.13	.74	.12	-7.06
3	4.42	2	4	2.21	.72	.24	-6.21
4	10.80	4	2	2.70	.32	.21	-7.40
5	.57	1	5	.57	.96	.16	-5.28
6	1.21	3	3	.40	.92	.46	-3.60
7	8.25	5	1	1.65	.48	.40	-5.12
null	16.00	6	0	2.67	.00	.00	-8.00

probably do better. Models 5 and 6 have ratios below 1.00, suggesting that the fits are a little too good (if one were testing an a priori hypothesis, the smaller the χ^2 the better, but if one were engaged in systematically freeing more parameters to achieve better fits, one might suspect that one had gone too far and was now just capitalizing on chance). The remaining model, model 7, fits reasonably well, given its degrees of freedom, and if one were to arrive at model 7 in the course of an exploration of model fitting, one might well choose to stop at this point. Notice the difference between this case and that of Table 2-12. If one were making a direct, theoretically motivated comparison of models 6 and 7, one would conclude that model 6 was better, but if one were just looking, one might suspect that the fit of model 6 was a little too good to be true--i.e, that it would be unlikely to do as well in a new sample--and prefer a model such as 7 that fits reasonably well with two fewer parameters.

Bentler and Bonett's normed fit index

Bentler and Bonett (1980) suggest that the goodness of fit of a particular model may be usefully assessed by placing it on a scale running from a perfect fit to the fit of some baseline "null model." Such a null model would be an arbitrary, highly restricted model--say, that all correlations are zero, or that all correlations are equal, or some such--which would represent a baseline level that any realistic model would be expected to exceed. The index would then represent the point at which the model being evaluated falls on a scale running from this null model to perfect fit. The normed fit index may be formally defined as:

$$\text{nfi} = (\chi_o^2 - \chi_k^2)/\chi_o^2,$$

where the subscripts k and o refer to the model in question and the null model, respectively.

Let us return to the example of Fig. 2.10 and Table 2-13. The next column of the table gives values of the normed fit index for the models of Fig. 2.10. Within a series of nested models, such as 1,2,3,4 or 1,5,6,7, the fit always gets poorer as fewer unknowns are fitted and the degrees of freedom increase. But the normed fit index gives some idea of how relatively good or poor the fit of a particular model might be. Thus, models 5 and 6 fit quite well, but models 4 or 7 do not constitute all that much of an improvement over the null model.

James, Mulaik, and Brett's parsimonious fit index

James, Mulaik, and Brett (1982) have suggested a modification of Bentler and Bonett's normed fit index, which they call the parsimonious fit index. This takes into account the number of degrees of freedom given up in order to arrive at any particular level of goodness of fit. Other things equal, a good fit attained by a parsimonious model, one with few unknowns (and hence more df), should represent the scientific ideal. James et al. suggest that if one were to modify Bentler and Bonett's index by multiplying it by the ratio of the degrees of freedom of the model under consideration to those of the null model, this would have the desired effect; that is:

$$\text{pfi} = (\text{df}_k/\text{df}_o) \text{ nfi}.$$

Values of the parsimonious fit index for the models of Fig. 2.10 are shown in the third column on the right of Table 2-13. Notice that within a nested series, such as 1,2,3,4,null or 1,5,6,7,null, instead of always increasing, as in the case of *nfi*, the values of *pfi* tend to increase and then decrease. The models with maximum values of *pfi* are the ones that best describe the data with the fewest unknown parameters to be solved for. The best model in this sense is model 6. Note that it scores much higher on the *pfi* than does model 5. The latter has a slightly better overall level of fit (*nfi* .96 vs. .92), but the fact that model 5 requires us to solve for 5 unknowns whereas model 6 gets by with only 3 gives the latter a decided advantage in its *pfi*. In fact, model 7, with only one unknown, comes out very well on the *pfi* ; its overall level of fit is not especially good (*nfi* = .48), but it consitutes a highly parsimonious explanation of the data.

Akaike's information criterion

A fourth criterion, similar in spirit to the parsimonious fit index, has been derived on the basis of information theory considerations by Akaike (see, for example, Cudeck & Browne, 1983). The derivation of Akaike's information criterion (AIC) assumes a χ^2 obtained via maximum likelihood, but presumably a GLS-based χ^2 could be used without serious problems. The AIC for a particular model is defined as the natural logarithm of the likelihood minus the number of free parameters (unknowns) solved for. The first, log likelihood term is readily obtained from the chi-square value as $-\chi^2/2$, therefore:

$$AIC = -\chi^2/2 - q,$$

where q parameters are solved for. AICs are thus always negative. The model fitting that provides the most information is that for which AIC is maximum, that is, closest to zero. In general, good fits--small χ^2s--from models with few unknowns lead to small negative AICs representing maximum parsimony, whereas large χ^2s from models with many free parameters yield large negative AICs representing lack of parsimony. The rightmost column of Table 2-13 shows the AIC values for the models of Fig. 2.10. Observe that AIC has a rough, though not perfect, correspondence with *pfi*; in particular, both criteria agree on model 6 as the most parsimonious fit to the data.

LISREL's indexes of fit

Jöreskog and Sörbom (1984) provide several goodness-of-fit indexes for LISREL that are based more or less directly on discrepancies between the observed covariance matrix **S** and the implied matrix **C**, rather than proceeding by way of χ^2 as do the indexes so far described.

The simplest of their measures is just the square root of the mean of the squared discrepancies between **S** and **C**, the root-mean-square residual, abbreviated RMR. This represents a kind of average of the absolute discrepancies between the observed and implied matrices; it is most easily interpretable when these are on a familiar scale, as would be the case if they were correlation matrices.

A slightly more sophisticated goodness-of-fit index GFI is based on a ratio of the sum of the squared discrepancies to the observed variances, thus allowing for scale. A version of GFI for maximum likelihood fitting is based on the sum of the squared discrepancies between the matrix product $C^{-1}S$ and the identity matrix **I**. The more similar **C** and **S** are, the smaller these discrepancies will be.

Finally, an adjusted index AGFI adjusts GFI by a ratio of the degrees of freedom of the restricted to the null matrix, much in the spirit of the James, Mulaik, and Brett parsimonious fit index.

Choice among the criteria

Here we are in the position of a person with seven watches--we may not know exactly what time it is, either, but if we consult all our watches and get a consensus, we should be pretty close. Actually, the situation is not quite analogous because not all the criteria are intended to measure exactly the same thing, but it is still the case that by consulting several we should have a more accurate picture of the situation.

One of the first four criteria, *nfi*, is a measure of goodness of fit per se--it is basically a rescaling of χ^2 onto a 0 to 1.0 scale. The other three, AIC, *pfi*, and χ^2/df, are indices of parsimonious goodness of fit, taking into account the number of free parameters required in order to achieve a given level of fit expressed as a χ^2. For each, as χ^2 increases for a fixed df, the fit gets worse, and as the number of free parameters decreases for fixed χ^2 it gets better. However, because some of the criteria involve sums and some ratios, the numerical details differ. As all are easily computed, it is perfectly feasible to obtain and consider several of them in making interpretations and decisions during exploratory model fitting.

Some of the Jöreskog and Sörbom criteria are also fairly easily computed directly, and in any case are automatically supplied if one is using LISREL. Again, the choice between an index like AGFI or one like RMR or GFI is a matter of whether one wishes to take degrees of freedom into account in evaluating the fit of a model to data.

Correlations Versus Covariances in Model Fitting

Earlier, in discussing the use of standardized or unstandardized variables in path models, it was noted that one often had a choice as to which scaling was used. What are some of the bases of such a choice?

There are several issues. First, the statistical theory underlying maximum likelihood and generalized least squares solutions has been developed for the case of covariance matrices rather than correlation matrices. Although workers in the area rather freely assume in practice that correlation matrices can be substituted for covariance matrices without serious risk, statistical purists are happier with covariance matrices. (Because correlation matrices *are* covariance matrices--of standardized variables--one might wonder why a problem arises. The answer is that statistical constraints are introduced when the same sample is used to standardize the variables and calculate the covariances.)

That is one side of the question, then: that in operating with a maximum likelihood or similar criterion, one may be on solider statistical ground if one is using covariances.

The other side of the question is that minimization programs may not work well if variables are on drastically different scales, a fact that would seem to commend the use of standardized variables, and thus correlations. Consider, say, years of education and annual income in dollars. If our minimization program works strictly with the raw units it will consider that a year of education is on a par with a dollar of annual income--which is probably *not* the attitude we want it to have when it is comparing the fit of alternative sets of trial values.

Such a program could spend all its effort in fitting annual incomes to within 10 dollars before even getting around to looking at discrepancies of "only" 6 or 7 years of education. Naturally, one can improve the situation by doing some rescaling of variables--by using, say, income in thousands rather than in dollars. But using correlations takes care of this automatically, and in a way that is responsive to the observed ranges of the variables, not just the sizes of the numbers.

There are other important issues to consider in making a decision whether or not to standardize. Two matters may especially be noted. First, if you are model fitting in several groups, covariances may often be the better choice--in calculating correlations separately in each group, important between-group information can be lost. If standardization is desirable on other grounds, you should consider doing it across the combined groups. And second, if the model to be fit involves equating one path coefficient or residual variance to another, you must decide whether it is on the standardized or unstandardized scale that you would expect the equality to hold. Equating on one will usually *not* be equating on the other. Most often, perhaps, the decision will go in favor of the unstandardized coefficient in such cases, but no flat rule can be prescribed.

Do it Yourself?

Before leaving the topic of model-fitting, we might consider one other option that a would-be modeler has: Do it your own way. If you have at least modest computer programming skills, it is quite feasible to build a practical IPSOL-like program by replacing its MINIM routine with a general-purpose search subroutine. As mentioned earlier, there are a number of such search programs around. One serviceable procedure is described in programming-level detail by Davidon (1975); there are a couple of minimization routines in the International Mathematical and Statistical Library of computer subroutines (IMSL, 1984), available at many computer centers; and other programs exist.

Why would someone want to do this? The most likely reason is to allow one to deal with models outside LISREL's scope. IPSOL, although a much less sophisticated program than LISREL, is more powerful in one important sense: Any desired relations among parameters and between parameters and observations can be written into subroutine FUNCT. LISREL can (in principle) represent any legal path diagram, but it is quite limited in the constraints it can place on the paths to be solved for. Essentially, one can specify that certain paths have fixed values, or are to be equal to others, and that's about it. Now this permits dealing with a wide range of important cases, but sometimes one would like to do more: to specify that the value of one path stands in some other numerical relation to another-- that it is one-half, or twice as large, or equal to its logarithm or its square, or that path *c* equals the product of paths *a* and *b*, or the like. With a general-purpose program you can do this; with a specialist program like LISREL you cannot--some other programs, such as COSAN (McDonald, 1978, 1980), offer more flexibility in this regard.

Most users, most of the time, should find LISREL or one of its fellows quite satisfactory. But if you are faced with frequent or extensive model-fitting analyses outside the scope of LISREL or whatever other standard programs you have available to you, it may be feasible to put together a package of your own.

Chapter 2 Notes

Search methods. Various procedures go by such names as steepest descent, Fletcher-Powell, Newton-Raphson, the EM algorithm, etc. They are usually discussed in numerical analysis texts under methods for solving sets of nonlinear simultaneous equations. LISREL uses a Fletcher-Powell variant. Schoenberg and Richtand (1984) discuss the use of the EM method for estimating factor analysis and measurement models.

IPSOL. For serious use, one would probably want to change the .001s in subroutine FUNCT at least to .0001s, and increase MAXFN.

Matrix formulations. McArdle and McDonald (1984) discuss some relationships among such formulations, and Bentler and Weeks (1985) comment.

Goodness-of-fit criteria. There has been recent interest in a fourth criterion, Browne's (1984) asymptotically distribution-free (ADF) criterion. For an application, see Huba and Bentler (1983); for an empirical comparison of all four criteria, see Huba and Harlow (1983). ADF, like ML and GLS, yields a χ^2 value for goodness of fit but makes weaker distributional assumptions than they do. Mooijaart (1985) provides some suggestions regarding computational efficiency.

Fit indexes. Tanaka and Huba (1985) point out that the two variants of Jöreskog and Sörbom's GFI are members of a general class of such indexes: They provide versions for generalized least squares and Browne's ADF (see preceding note). Other recent discussions of goodness-of-fit indexes may be found in Hoelter (1983) and Sobel and Bohrnstedt (1985).

Standardization across populations. How LISREL does it is discussed by Acock and Fuller (1985).

Other model-fitting approaches. As a variation on do-it- yourself, Lee and Jennrich (1984) point out how one can use a general nonlinear least squares program, PAR in the BMDP series, to achieve very flexible model fitting. The procedure runs somewhat more slowly than LISREL and requires a modest amount of FORTRAN-type programming, but it is much more general in the kind of constraints that may be imposed on the variables being solved for. As an opposite strategy, Rindskopf (1983, 1984b) points out that by introducing "phantom" variables and other devices LISREL may be tricked into imposing constraints that are not normally within its repertoire (e.g., that one path exceed another, or that all residual variances be positive, etc.).

Chapter 2 Exercises

In problems 1, 5, 8, and 9, at the instructor's discretion, other methods or programs may be used, or you may be asked merely to set up the problem, not to solve it numerically.

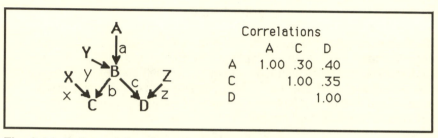

	Correlations		
	A	C	D
A	1.00	.30	.40
C		1.00	.35
D			1.00

Fig. 2.11 Path model and correlations for Problem 1.

1. Use IPSOL to solve the path model of Fig. 2.11, with the correlations shown to its right. Reproduce the implied correlations from your solution as a check. Solve also for the residual paths.

2. Draw a cross-section of a perverse terrain in which very good solutions exist but a search program along the lines of IPSOL would have almost no chance of finding them.

3. Set up the three McArdle-McDonald matrices for the path diagram of Problem 1.

4. An investigator believes that ambition, of which he has three measures, is a cause of achievement, for which he has two measures. If he sets this up as a normal LISREL problem, what would be the dimensions of the following matrices: **PH, BE, PS, LX, GA, LY, TD**, and **TE**?

5. In the study of Problem 4, which used 60 subjects, the following correlations were observed. Use LISREL to solve for the (standardized) path values, with a maximum likelihood criterion. Interpret the results.

	Ach1	Ach2	Amb1	Amb2	Amb3
Ach1	1.00				
Ach2	.60	1.00			
Amb1	.30	.20	1.00		
Amb2	.20	.30	.70	1.00	
Amb3	.20	.10	.60	.50	1.00

6. Four nested models based on a 4 x 4 variance-covariance matrix have 3, 5, 6, and 9 unknowns and yield χ^2s of 16.21, 8.12, 2.50, and 0.08, respectively. What conclusions about models or the differences between models could you draw at the .05 level of significance?

7. Compute the four χ^2-based criteria for exploratory model fitting for the data of Problem 6, assuming a null model with no free parameters and $\chi^2 =$ 25.00. What interpretation would you make?

8. Solve the path model of Problem 1 using LISREL and a least squares criterion. Compare the results. (Hint: One can set the problem up along the lines of Fig. 2.12, in which path *b* is fixed to 1.0 and the observed independent variable A is treated as a latent variable measured without error. Obtain a standardized as well as the unstandardized solution. Remember that X variables come last in LISREL input!)

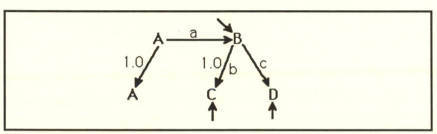

Fig. 2.12 Hint for Problem 8.

9. Convert the unstandardized solution of Problem 8 to the standardized solution, using the rules for standardized and unstandardized path coefficients from Chapter 1.

Chapter Three:
Varieties of Path and Structural Models--I

In this chapter and the next we consider a number of different applications of path and structural models. This serves two purposes. First, it gives some sense of various ways in which these methods can be employed, as well as practice in applying and interpreting them. Second, it introduces some additional concepts and techniques useful in dealing with path and structural models, both in general and in some important special cases.

Structural and Measurement Models

As we have seen, Jöreskog makes a useful distinction between two conceptually distinct parts of many complex path models, namely, a *structural* part and a *measurement* part. The structural part of a model specifies the relationships among the latent variables, and the measurement part specifies the relationship of the latent to the observed variables.

An example from a desegregation study

Figure 3.1 gives an example. This is a path diagram of part of a study of school desegregation. The diagram adopts a convention fairly common among users of structural models, of representing latent variables by circles or ovals, and observed variables by squares or rectangles. This is helpful in keeping things straight in complicated models.

There are five latent variables, listed beside the diagram. Each is indexed by two or three observed variables, identified below the diagram. Collectively, these constitute the measurement model, shown in the top part of the next figure, Fig. 3.2. The structural model consists of the relationships among the five latent variables, shown at the bottom of Fig. 3.2.

Note that the structural and measurement models play rather distinct roles in the overall path model. One could very easily alter either without changing the other. Thus, one might maintain the same structural model of relationships among the latent variables but change the measurement model by using

76

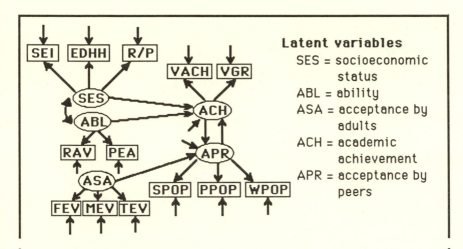

Latent variables
  SES = socioeconomic
        status
  ABL = ability
  ASA = acceptance by
        adults
  ACH = academic
        achievement
  APR = acceptance by
        peers

Observed variables
  SEI = socioeconomic index
  EDHH = education of head of household
  R/P = rooms per person in house
  VACH = verbal achievement score
  VGR = verbal grades
  RAV = Raven's Progressive Matrices Test
  PEA = Peabody Picture Vocabulary Test
  FEV = father's evaluation
  MEV = mother's evaluation
  TEV = teacher's evaluation
  SPOP = seating popularity
  PPOP = playground popularity
  WPOP = schoolwork popularity

Fig. 3.1 Path model used in a desegregation study (Maruyama & McGarvey, 1980).

different tests or measurements to index the latent variables. Alternatively, one could keep the same measures but change the structural model by making different assumptions about the relationships among the latent variables--one could assume, say, that the child's academic achievement influences peer approval but not vice versa, or that acceptance by adults is affected by the child's academic achievement.

The measurement model is a variant of confirmatory factor analysis. One could consider the top part of Fig. 3.2 to consist of five small Spearman

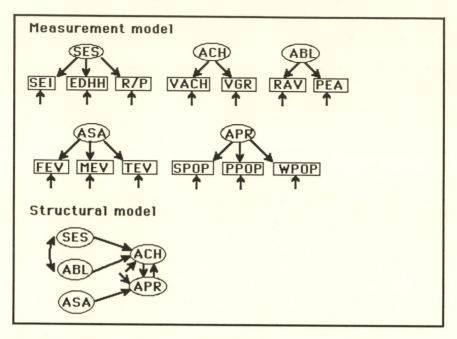

Fig. 3.2 Measurement and structural components of path model of Fig. 3.1.

general-factor problems: In each case the latent variable can be seen as a factor general to the tests that measure it, with the residual arrows--shown here below the squares--representing influences specific to the tests. This would not, however, be an altogether satisfactory way to proceed, because it is not only the correlations of, say, SEI, EDHH, and R/P among themselves that provide information concerning the paths between them and SES; such information is also supplied in the relative correlations of these variables with others in the model. In fact, a direct Spearman approach would not work at all for the latent variables with only two indicators, which require at least some additional correlations in order to obtain a unique solution. It is perhaps more appropriate, therefore, to think of the measurement model as a single, multiple-factor confirmatory factor analysis. So far as the measurement model is concerned, the relationships among these factors are unanalyzed correlations. It is the structural model that interprets these correlations as resulting from a particular set of causal relationships among the latent variables.

In practice, the usual procedure is to solve the measurement and structural models simultaneously because, in so doing, one brings to bear all information available about each path.

A solution of the model

Maruyama and McGarvey present correlations among the 13 observed variables for a sample of 249 children--the correlations given in Table 3-1. One could in principle write path expressions for each of these 78 correlations in terms of the unknown paths and solve them for the path values, but Maruyama and McGarvey preferred to set up the appropriate LISREL matrices and let the program handle the details.

Table 3-1 Correlations among observed variables in desegregation study (data from Maruyama & McGarvey, 1980), N = 249

	SEI	EDH	R/P	VACH	VGR	RAV	PEA	FEV	MEV	TEV	SP	PP	WP
SEI	1.00												
EDHH	.56	1.00.											
R/P	.17	.10	1.00										
VACH	.17	.30	.19	1.00									
VGR	.16	.21	-.04	.50	1.00								
RAV	.06	.15	-.00	.29	.28	1.00							
PEA	.16	.21	.28	.40	.19	.32	1.00						
FEV	.01	-.04	-.04	.01	.12	.10	-.06	1.00					
MEV	-.07	-.05	.00	.13	.27	.16	-.07	.42	1.00				
TEV	-.02	-.01	.04	.21	.27	.14	.08	.18	.31	1.00			
SPOP	.05	.04	.02	.28	.24	.08	.13	.07	.15	.25	1.00		
PPOP	.10	.10	-.04	.23	.18	.09	.17	.02	.08	.08	.59	1.00	
WPOP	.10	.17	-.03	.32	.40	.14	.17	.08	.17	.33	.55	.49	1.00

The results are shown in Figure 3.3, in which all latent variables are standardized, so that these numbers are ordinary path coefficients and correlations.

The χ^2 for testing the goodness of fit of the model was 138.55, based on 59 degrees of freedom. A χ^2 of 138.55 based on 59 df is not very likely (p < .01) to occur with a sample of this size if the model of Fig. 3.1 truly holds in the population. Thus, we may reject the model. The sample size of 249 is not so huge that any trivial departure from the model would lead to its rejection, so we should probably take the negative outcome seriously. The ratio of χ^2 to df, 2.35, suggests that a better fitting version of the model might be found. However, Maruyama and McGarvey did not pursue matters further at this point.

One feature of this model, the reciprocal paths between ACH and APR, represents a step beyond the models that we have so far considered. We look further at such looped models in the next chapter; for the moment we need merely note that with adequate data their solution in LISREL presents no

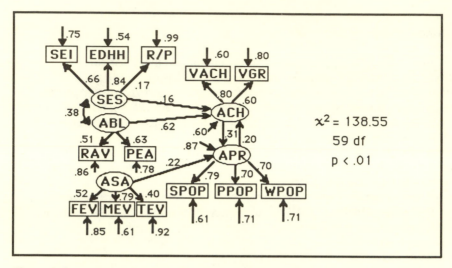

Fig. 3.3 Standardized solution to path diagram of Fig. 3.1 from correlations of Table 3-1 (after Maruyama & McGarvey, 1980).

special problems.

    We need not go into all the details of Maruyama and McGarvey's solution procedure, but a few points may be helpful. First, as is usual with LISREL, the correlation matrix was treated as if it were a variance-covariance matrix, with $n(n + 1)/2 = 91$ distinct observed values--the 13 variances plus the 78 covariances on one side of the diagonal. Maruyama and McGarvey employed a mixed technique for scaling the latent variables. On the independent variable side they specified standardized latent variables, leaving 16 unknowns to be solved for in the measurement model: the $3 + 2 + 3 = 8$ paths from the latent variables SES, ABL, and ASA to the eight manifest variables measuring them, plus the corresponding 8 residual paths. On the dependent variable side of the measurement model, they fixed one path from each of the latent variables to 1.0, leaving only $1 + 2 = 3$ paths to be solved, plus the 5 residual paths, or 8 unknowns. Altogether, then, there were a total of $16 + 8 = 24$ unknown paths to be solved for in the measurement model.

    In the structural model there were a total of 8 unknowns: one correlation among independent latent variables, three paths from independent to dependent latent variables, two reciprocal paths between the latter, and two residuals for the downstream latent variables. Thus, there were altogether $24 + 8 = 32$ unknowns to be solved for; and 91 observed values minus 32 unknowns yields the 59 df.

    The initial solution provided by LISREL (using a maximum likelihood criterion) was thus standardized on the independent variable side but not on the dependent variable side; however, Maruyama and McGarvey also obtained a fully standardized solution, and it is that which is reported in Fig. 3.3.

Confirmatory Factor Analysis

Traditionally, a latent variable analysis that is called a *confirmatory factor analysis* is one in which a hypothesized model is fit to a correlation matrix, the model being mainly a measurement model with the structural model confined to simple correlations among the latent variables. The restriction to correlation matrices, though usual, is not essential--covariance matrices can be factor analyzed, and sometimes are.

The main result of a factor analysis, as noted in Chapter 1, is a table showing the *factor pattern*, or values of the paths between the latent and observed variables. If the latent variables are correlated, there will also be a table of their intercorrelations (*factor intercorrelation matrix*). In the case of correlated factors, there may also be reported a table of the correlations between observed and latent variables (*factor structure matrix*).

A study of attitudes toward police

In Chapter 1 we considered some simple artificial examples of confirmatory factor analysis. Here we look at a case (McIver, Carmines & Zeller, 1980) in which several hypotheses were fit to correlations based on a large real-life data set.

The correlations, from a study of attitudes toward police, are given in Table 3-2. They are based on telephone interviews with a total of some 11,000

Table 3-2 Correlations among nine items in police survey (data from McIver, Carmines, & Zeller, 1980)

	1	2	3	4	5	6	7	8	9
1. Police service	1.00	.50	.41	.33	.28	.30	-.24	-.23	-.20
2. Responsiveness	.48	1.00	.35	.29	.26	.27	-.19	-.19	-.18
3. Response time	.42	.37	1.00	.30	.27	.29	-.17	-.16	-.14
4. Honesty	.34	.30	.26	1.00	.52	.48	-.13	-.11	-.15
5. Courtesy	.31	.27	.24	.51	1.00	.44	-.11	-.09	-.10
6. Equal treatment	.29	.26	.23	.48	.44	1.00	-.15	-.13	-.13
7. Burglary	-.24	-.21	-.19	-.14	-.13	-.13	1.00	.58	.47
8. Vandalism	-.22	-.19	-.17	-.13	-.12	-.11	.58	1.00	.42
9. Robbery	-.18	-.16	-.14	-.11	-.10	-.09	.47	.42	1.00

Note: original correlations are above diagonal; those implied by 3-factor solution are below it.

respondents in 60 neighborhoods in 3 U.S. metropolitan areas. Included in the table are the intercorrelations among 6 items reflecting attitudes toward the quality of police services, plus 3 items having to do with the likelihood of burglary, vandalism, and robbery in the neighborhood.

The authors had originally surmised that the six attitude items might form a single general dimension of attitude toward police, and indeed they are all mutually positively intercorrelated in Table 3-2. But inspection of the table suggested that there might be two distinct subclasses of attitude items, judging from somewhat higher correlations within than across item subsets. The first three items, having to do with the general quality of police services, their responsiveness to citizen needs, and the rapidity with which the police answered a call, seemed to go together, as did the second three, having to do with the personal qualities of the police--their honesty, courtesy, and fairness. The three items concerning likelihood of various kinds of crime also seemed to group together and to be mildly negatively correlated with the first six items having to do with the perceived quality of the police service.

Consequently, a hypothesis of three correlated factors was fit to the data. (The fact that this hypothesis was arrived at on the basis of preliminary inspection of the data means that the chi-square tests should not be regarded as yielding strict probabilities, but rather as more informal indices of goodness of fit.)

Table 3-3 gives the results of fitting the hypothetical three-factor model shown in Fig. 3.4. Each item has a substantial loading on its corresponding factor. The first two factors of police attitudes are substantially correlated (r=.62), and the third, crime factor, is negatively correlated with both, somewhat more highly with the first factor (police service) than with the second (personal qualities of the police). The communalities (h^2), which in this case are simply

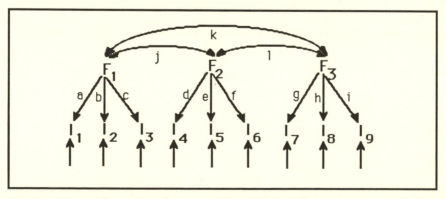

Fig. 3.4 Simple 3-factor model for data of Table 3-2 from survey on attitudes toward police (McIver et al., 1980).

equal to the squares of the individual paths *a, b, c,* etc. (see Fig. 3.4 for the reason), suggest that only around half the item variances are being accounted for by the common factors, with the rest presumably due to specific factors and error.

The three common factors were assumed to be in standard-score form. Forty-five observed variances and covariances (9 x 10 / 2) were fit using 21 unknowns (*a* through *l*, in Fig. 3.4, plus the 9 residuals), leaving 24 df for the χ^2 test. The obtained χ^2, with this huge sample, is a highly significant 225.10. Nevertheless, the solution does a fairly good job of accounting for the data. Correlations implied by the solution of Table 3-3 are shown in Table 3-2, below the principal diagonal. The largest discrepancy between observed and expected correlations is .06, and the bulk of the differences are .02 or less. For many purposes one might be perfectly content with this good a fit.

However, the discrepancies of .04, .03, and .06 of item 3 (police response time) with items 4 through 6 (personal qualities) suggest that the fit might be improved a little if F_2 as well as F_1 were allowed to influence item 3. Such a

Table 3-3 Factor pattern and factor intercorrelations for model of Fig. 3.4 (McIver et al., 1980)

	Factor pattern			
Item	F_1	F_2	F_3	h^2
1. Police service	.74	.00	.00	.55
2. Responsiveness	.65	.00	.00	.42
3. Response time	.57	.00	.00	.32
4. Honesty	.00	.75	.00	.56
5. Courtesy	.00	.68	.00	.46
6. Equal treatment	.00	.64	.00	.41
7. Burglary	.00	.00	.80	.64
8. Vandalism	.00	.00	.72	.52
9. Robbery	.00	.00	.59	.35

Factor intercorrelations			
	F_1	F_2	F_3
F_1	1.00	.62	-.41
F_2		1.00	-.24
F_3			1.00

$\chi^2 = 225.10$, 24 df, p < .001

solution yielded a reduction of χ^2 from 225.10 to 128.00, at the cost of one degree of freedom--a very substantial improvement in goodness of fit. Substantively, however, the change makes little difference except in the paths to item 3 (the path from F_1 drops from .57 to .45 as that from F_2 rises from 0 to .15; the estimated correlation of F_1 and F_2 also drops slightly, from .62 to .59).

The authors went on to test models allowing the residual variances to be correlated for a couple of pairs of variables, 4 and 9, and 2 and 7, further reducing the χ^2 to 90.33. This is still a highly significant improvement, if one takes the χ^2s seriously, but as this amounts to introducing new factors to explain the discrepancies of single correlations, it is not very helpful from the point of view of parsimony.

The overall χ^2 is still highly significant with 21 df, despite the fact that at this stage the data are being fit very much ad hoc. With very large samples one needs to be careful not to confuse statistical with practical significance. Some of the small deviations of the data from the model may indeed not be due to chance, but to introduce a hypothetical variable to account for each one of them is unlikely to be of much value for either science or practice.

Some Psychometric Applications of Path and Structural Models

A number of problems in psychometrics lend themselves to latent variable methods. We considered an example involving test reliability in Chapter 1. In the present section, we look at examples involving parallel and congeneric tests, and the separation of trait and method variance.

Parallel and congeneric tests

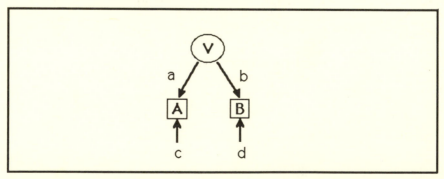

Fig. 3.5 Parallel or congeneric tests.

First, some definitions: Two tests are said by psychometricians to be *parallel* if they share equal amounts of a common factor, and each also has the same amount of specific variance. Consider Figure 3.5. If tests A and B are parallel, *a* and *b* would be equal, and so would *c* and *d*. V would represent the common factor the two tests share.

Tests are said to be *congeneric* if they share a common factor, but not necessarily to the same degree. Tests A and B would still be congeneric--because they share V--even though *a* were not equal to *b*, nor *c* to *d*. (We encountered also in Chapter 2 a third, intermediate condition, *tau-equivalence*, in which *a = b*, but *c ≠ d*; however, this will not be involved in the present example.)

Jöreskog provides a structural analysis of some data gathered by F.M. Lord on four vocabulary tests. Tests A and B were short tests given under leisurely conditions, whereas C and D were longer tests given under time pressure. The variance-covariance matrix of the four tests is given in Table 3-4.

Jöreskog carried out tests of four hypotheses, which can be expressed in terms of Fig. 3.6 by imposing the conditions noted in parentheses:

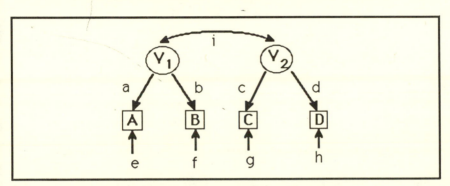

Fig. 3.6 Hypotheses about parallel and congeneric tests.

H1: Tests A and B are parallel, as are C and D, and all four tests are congeneric. (*a = b*, *e = f*; *c = d*, *g = h*; the correlation *i = 1.0*--that is, V_1 and V_2 are identical except possibly for scale.)

H2: Both test pairs are parallel, as in H1, but the two pairs are not necessarily congeneric. (*a = b*, *e = f*; *c = d*, *g = h*.)

H3: All four tests are congeneric but are not necessarily parallel. (*i = 1.0*.)

H4: A and B are congeneric, as are C and D, but the two pairs need not be congeneric with each other. (Fig. 3.6 as it stands.)

Note that these hypotheses form two nested series, H1, H2, H4, and H1, H3, H4, within which χ^2 comparisons may be made.

Table 3-4 Covariance matrix for four vocabulary tests (data from Lord; Jöreskog an Sörbom, 1979, p. 55), N = 649

Test	A	B	C	D
A	86.40			
B	57.78	86.26		
C	56.87	59.32	97.28	
D	58.90	59.67	73.82	97.82

Table 3-5 Hypothesis tests for problem of Fig. 3.5 with data of Table 3-4 (after Jöreskog & Sörbom, 1979, p. 55)

Model	Unknowns	χ^2	df	p
H1	4	37.33	6	<.01
H2	5	1.93	5	.86
H3	8	36.21	2	<.01
H4	9	.70	1	.70

Model comparison	χ^2_{diff}	df	p
H2 vs H1	35.40	1	<.01
H4 vs H3	35.51	1	<.01
H3 vs H1	1.12	4	>.80
H4 vs H2	1.23	4	>.80

Table 3-5 shows the results of several χ^2 tests. In the upper part of the table the models representing the four hypotheses are tested individually. Models H1 and H3 can be rejected; models H2 and H4 cannot. The rejected models H1 and H3 contain the assumption that all four tests are congeneric, whereas models H2 and H4 do not contain this assumption.

The specific comparisons of H2 versus H1 and H4 versus H3 (bottom part of Table 3-5) represent pairs of models that are equivalent except for the assumption that tests A and B are congeneric with tests C and D. Both comparisons show that this assumption is not tenable. The other two comparisons shown, H3 versus H1, and H4 versus H2, test the assumptions that A is parallel to B and C is parallel to D. These assumptions remain quite tenable.

Note that the sample size is fairly large (N = 649). Although A and B have been shown to be noncongeneric with C and D by large χ^2s, the two pairs of tests are not in fact *very* different. The correlation *i* was estimated in the solution to model H2 as approximately .90, suggesting that although the speeded and unspeeded vocabulary tests are not measuring quite the same thing, what they measure does not differ much for practical purposes.

Fitting one of these psychometric models is equivalent to carrying out a confirmatory factor analysis. Another psychometric model that shares this character is the multitrait-multimethod model, to which we now turn.

Multitrait-multimethod models

The multitrait-multimethod model (Campbell & Fiske,1959) is an approach to psychological measurement that attempts to separate out true variance on psychological traits from variance due to measurement methods. The basic strategy is to measure each of several traits by each of several methods. Correlations among these measurements are arranged in a multitrait-multimethod matrix that enables one to assess *convergent validity*, the tendency for different measurement operations to converge on the same underlying trait, and *discriminant validity*, the ability to discriminate among different traits.

Table 3-6 is an example of a multitrait-multimethod correlation matrix. It is based on part of a study by Bentler and McClain (1976) in which 68 fifth-grade girls were measured in each of three ways on four personality variables: impulsivity, extraversion, academic achievement motivation, and test anxiety.

For the self-rating measure, each of the girls filled out four standard personality questionnaires, one for each trait. For the teacher ratings, teachers were asked to rank the children in their class on each of the four variables under consideration. Their ratings were converted to scores using a normalizing transformation.

The peer ratings were obtained by a sociometric procedure, in which children in the class were asked to write the names of children who fit various descriptions. Four to eight items were used per trait. An example of an item for extraversion was: "Which children like to be with other children a lot?"

The off-diagonal elements in a multitrait-multimethod matrix such as Table 3-6 can be classified into three groups. In the triangles adjacent to the main diagonal are *within-method, cross-trait* correlations. They are underlined in Table 3-6. An example would be the .42 at the start of the third row, the correlation between extraversion and impulsivity, both measured by peer ratings. In the diagonals of the square blocks in the rest of the table, given in boldface type, are the *within-trait, cross-method* correlations, also known as the *validity diagonals*. An example would be the .64 at the start of the fifth row, the correlation between peer and trait assessments of extraversion. The remaining, unmarked off-diagonal elements are the *cross-trait, cross-method* correlations.

Table 3-6 Multitrait-multimethod correlation matrix for four traits measured by peer, teacher, and self-ratings (data from Bentler & Lee, 1979), N = 68

Trait and method of measurement

	Ep	Ap	Ip	Mp	Et	At	It	Mt	Es	As	Is	Ms
Ep	1.00											
Ap	-.38	1.00										
Ip	.42	-.21	1.00									
Mp	-.25	.54	-.54	1.00								
Et	.64	-.15	.26	-.05	1.00							
At	-.29	.66	-.19	.44	-.25	1.00						
It	.38	-.09	.56	-.19	.59	-.14	1.00					
Mt	-.22	.51	-.33	.66	.06	.62	-.05	1.00				
Es	.45	-.05	.12	.10	.50	-.05	.36	.17	1.00			
As	.04	.38	-.03	.14	.08	.30	.09	.16	.02	1.00		
Is	.33	-.13	.35	-.18	.41	-.14	.45	-.13	.43	.16	1.00	
Ms	-.21	.37	-.44	.58	-.01	.41	-.10	.62	.06	.04	-.37	1.00

Note: Trait: E = extraversion, A = test anxiety, I = impulsivity, M = academic achievement motivation. Rater: p = peer, t = teacher, s = self. Correlations: underlined = within-method, cross-trait; boldface = within-trait, cross-method.

High correlations in the validity diagonals are evidence of *convergent validity,* the agreement of different methods of measuring the same trait. Low correlations elsewhere provide evidence of *discriminant validity,* that the putatively different traits really are distinct. Within-method, cross-trait correlations in excess of cross-method, cross-trait correlations are evidence of the presence of *method variance,* associations among measures stemming from properties of the measurement methods used.

In Table 3-6 the correlations in the validity diagonals are generally positive and appreciable in size (.30 to .66, with a mean of .51), suggesting reasonable convergent validity. They tend to be decidedly higher than the cross-method, cross-trait correlations (mean absolute value of .20), indicating some degree of discriminant validity. However, the latter correlations are by no means always negligible (they range up to about .50), suggesting some overlap among the traits. The within-method, cross-trait correlations (mean absolute value of .28) are slightly higher than the cross-method, cross-trait correlations; thus there appears to be some method variance.

Such a multitrait-multimethod matrix can be represented by a path model in the manner shown in Figure 3.7. The 12 observed variables are shown in the center row of the figure. Four latent variables representing true scores on

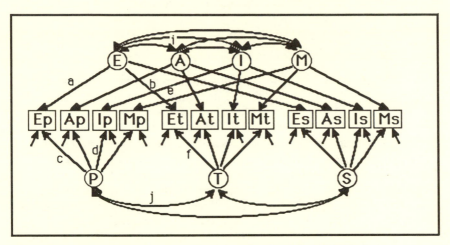

Fig. 3.7 Path model of multitrait-multimethod matrix of Table 3-6.

the four traits are shown in the circles at the top of the diagram. Three latent variables representing the effects of the three methods are shown in the circles at the bottom. Each observed measurement is determined by a trait and a method (e.g., arrows *a* and *c* for *Ep*), plus a residual. The traits may be intercorrelated--for example, extraversion and impulsivity might be related (arrow *i*). So might the methods--for example, peer and teacher ratings (arrow *j*). However, it is assumed in this particular diagram that the design of the experiment has insured that there will be no systematic correlations between methods and traits (no curved arrows connecting the top and bottom circles).

Within-trait, cross-method correlations are produced by direct paths via the trait in question, and possibly by indirect paths via the correlations among methods. For example, the correlation between peer and teacher ratings of extraversion may be expressed as:

$$r_{EpEt} = ab + cjf.$$

Within-method, cross-trait correlations are the other way around-- direct paths via methods and indirect paths via possible correlations among traits. For example, the correlation between peer ratings of extraversion and impulsivity may be expressed as:

$$r_{EpIp} = cd + aie.$$

Cross-method, cross-trait correlations are produced only via indirect paths of both kinds. For example:

$$r_{IpEt} = eib + djf.$$

The model of Fig. 3.7 involves 12 paths from traits to measures, 12 paths from methods to measures, 12 residuals, 6 intercorrelations among traits, and 3 intercorrelations among measures, a total of 45 unknowns to be estimated from 78 observed variances and covariances (12 x 13 / 2), leaving 33 degrees of freedom for a χ^2 goodness-of-fit test. (This assumes standardized factors: Alternatively, one could set 7 paths, 1 per factor, to 1.0, and solve instead for the 7 factor variances, leaving the number of df unchanged at 33.)

A solution of the path model by Bentler and Lee (1979) is given in Table 3-7. The values of the 12 paths from the trait factors to the measurements are given in the first subtable. Obviously, the measurements are substantially determined by the traits--somewhat more so for the peer and teacher ratings than for the self-ratings.

Determination of the measurements by the methods is shown in the second subtable. On the whole, these numbers are a bit lower than those in the first part of the table, but they are by no means all low--measurements of impulsivity, for example, seem to be about as much determined by methods (third column of second table) as by the trait (third row of first table).

The trait factors are substantially interrelated (third subtable). Test anxiety and academic achievement motivation tend to go together, as do extraversion and impulsivity, with the two pairs negatively related to each other. The method factors (fourth subtable) are fairly independent of one another, except for a modest correlation betwen teacher and self-ratings.

Bentler and Lee fit their model using a generalized least squares criterion, obtaining a χ^2 of 43.88 with 35 df, indicating a tolerable fit (p > .10; χ^2/df = 1.25). They had 35 df rather than 33 because they additionally set the unique variances of two variables, Mp and Is, to zero. (This was done to forestall a tendency for these variances to go negative during the solution, an awkward event known in factor analytic circles as a "Heywood case.") Negative variances of any kind are, of course, not plausible. Unfortunately, however, neither are empirical measures with no unique variance because this implies, among other things, the absence of errors of measurement. Thus, the present solution, although an adequate fit statistically, cannot be regarded as entirely satisfactory. (Actually, Bentler and Lee fit several other hypothetical models to these data as well--the interested reader can consult their article for details.)

Table 3-7 Solution to model of Fig. 3.7 for data of Table 3-6 (after Bentler & Lee, 1979, Table 7)

Trait factors	Peer	Ratings Teacher	Self	
Extraversion	.98	.62	.42	
Anxiety	.77	.91	.35	
Impulsivity	.78	.64	.42	
Motivation	.72	.90	.66	
Method factors	E	Traits A	I	M
Peer ratings	.15	-.25	.32	-.68
Teacher ratings	.74	-.19	.49	.13
Self ratings	.34	.17	.89	-.22

Trait factor intercorrelations

	E	A	I	M
Extraversion	1.00	-.35	.52	-.24
Anxiety		1.00	-.26	.74
Impulsivity			1.00	-.48
Motivation				1.00

Method factor intercorrelations

	P	T	S
Peer ratings	1.00	.08	.04
Teacher ratings		1.00	.32
Self ratings			1.00

Testing the relationship between trait and method variance

The previous section, on multitrait-multimethod matrices, showed how latent variable analysis could be used to assess the relative contribution of trait scores and measurement methods to observed scores. In this section we consider a related problem: whether, for a particular questionnaire scale, a

91

certain artifact occurs, and if it does, whether it tends to be correlated with the true score.

The particular questionnaire scale considered is the Psychasthenia (*Pt*) scale of the MMPI. This is a scale that purports to measure certain tendencies toward anxiety, obsessive worrying, and lack of confidence. It happens, however, that most of the items on this scale are expressed in such a way that agreeing with them gives one a high score on *Pt*, and disagreeing with them gives one a low score. This raises an alternative possibility. Suppose there are people who are just generally inclined to agree to anything--call them "acquiescent." Given the nature of the *Pt* items, individuals with a strong general tendency toward acquiescence would also get high *Pt* scores. So what, in fact, does *Pt* measure--a trait of psychasthenia, a trait of acquiescence, or both?

One method that has been used to study this question is to reverse the wording of items and see how this affects responses. If a person says "true" to an item like "I worry a great deal," you don't know whether he is a worrier or an agreer. But if you later give him the item "I worry very little," his response will give you a clue. If he says "true" again, you judge him to be an agreer. If he says "false," you judge him to be a worrier.

Bramble and Wiley (1974) used a latent variable approach to address this question, analyzing data from original and reversed versions of the *Pt* scale that had been gathered earlier by Bock, Dicken, and Van Pelt (1969). A group of 81 undergraduates had taken the *Pt* items in their original and reversed forms on two occasions. Fig. 3.8 gives a path model representing this situation. Three latent variables are postulated: the true trait of psychasthenia, the trait of acquiescence, and a third latent variable representing change between occasions of testing.

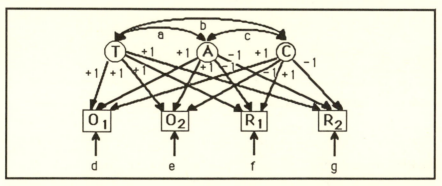

Fig. 3.8 Path model for original and reversed versions of an MMPI scale administered on two occasions. T = trait, A = acquiescence, C = temporal change; O = original items, R = reversed items; 1,2 = two occasions of testing.

92

As seen in the figure, Bramble and Wiley adopted a slightly different strategy than has been used in the examples so far, although it is still a variant of confirmatory factor analysis.

They set the paths from the latent to observed variables at fixed values to reflect the theoretical design of the experiment and solved for the variances of the latent variables T, A, and C, the covariances *a*, *b*, and *c* among them, and the specific variances *d*, *e*, *f*, and *g*, under various hypotheses.

The scales are assumed all to be keyed to measure the trait, so that a large amount of the true trait will always contribute positively to the obtained score: hence the positive signs on all the paths from T. A tendency toward acquiescence would work in the same direction for the original scales (paths of +1.0 from A to O_1 and O_2) but would work in the opposite direction for scales with reversed items (paths of -1.0 from A to R_1 and R_2). The signs of the paths from C are taken so that tests given on the same occasion will tend to have an increment in correlation (O_1 and R_1, O_2, and R_2), whereas tests given on different occasions will have lowered correlations (O_1 with R_2, O_2 with R_1).

Bramble and Wiley considered four models, based on the following hypotheses:

H1: T, A, and C are uncorrelated, and all residual variances are equal. ($a = b = c = 0$, $d = e = f = g$. 4 unknowns--variances T, A, C, and the residual variance; 6 df.)

H2: T, A, and C are uncorrelated, but residual variances may be unequal. ($a = b = c = 0$. 7 unknowns, 3 df.)

H3: Residual variances equal, but T, A, and C may be correlated. ($d = e = f = g$. 7 unknowns, 3 df.)

H4: T, A, and C correlated and unequal residual variances. (Fig. 3.8 as it stands. 10 unknowns, 0 df.)

Table 3-8 shows the results for *Pt*, based on the data of Bock et al. given

Table 3-8 χ^2 tests of various hypotheses, based on path model of Fig. 3.8 and data of Table 3-9

Model	Unknowns	χ^2	df	p
H1	4	14.63	6	<.05
H2	7	12.70	3	<.01
H3	7	2.64	3	>.30
H4	10	0.00	0	1.00
Comparisons		χ^2_{diff}	df	p
H1 vs H2		1.93	3	>.50
H1 vs H3		11.99	3	<.01

Table 3-9 Observed variance-covariance matrix for *Pt* scale, for path model of Fig. 3.8 (data from Bock, Dicken, & Van Pelt, 1969), N = 81

Scale version	O_1	O_2	R_1	R_2
Original, time 1	34.82			
Original, time 2	30.52	35.00		
Reversed, time 1	28.18	24.84	29.38	
Reversed, time 2	25.52	27.61	22.32	26.78

in Table 3-9. Models H1 and H2, which assume uncorrelated T, A, and C, do not fit the data; model H3, which assumes equal residual variances, does fit. (Model H4, with as many unknowns as observations, also fits the data, of course.)

The direct comparison of H3 and H1, identical except for the presence or absence of correlation among T, A, and C, shows that allowing T, A, and C to be correlated yields a statistically significant improvement in fit (p < .01).

Table 3-10 shows the values of the unknowns that gave the best fit for model H3, using a maximum likelihood criterion. Also shown is the

Table 3-10 Fitted values for model H3, and implied covariance matrix

Unknown	Estimate	Standardized
s^2_T	27.14	1.00
s^2_A	.62	1.00
s^2_C	1.36	1.00
cov_{TA}	1.88	.46
cov_{AC}	-.26	-.28
cov_{TC}	.22	.04
s^2_{resid}	2.36	

Implied variance-covariance matrix

	O_1	O_2	R_1	R_2
O_1	35.16			
O_2	30.16	35.32		
R_1	28.32	24.64	28.68	
R_2	25.68	27.44	22.64	26.76

standardized version of the solution, which provides the correlations among the three latent variables. At the bottom of the table is the variance-covariance matrix implied by the solution under H3.

From the implied variance-covariance matrix, we see that the model does a reasonable job of fitting the data--the largest absolute discrepancy from the observed data is .70 and the median is .26. From the standardized solution we see that there is a modest positive correlation of .46 estimated between the trait factor *Pt* and the method factor of acquiescence; that is, scores on psychasthenia could be mildly inflated by a tendency for those high on the trait also to be acquiescent. However, if one compares the estimated trait variance with the variance of O_1, which would represent the typical single administration of the original test, one finds that 77% of the variance (27.14/35.16) appears to be due to the trait, with only about 11% (2 x 1.88 / 35.16) due to the covariation of *Pt* and acquiescence, and 2% (.62/35.16) attributable to acquiescence as such.

Structural Models--Controlling Extraneous Variables

The technique of partial correlation, and the related method of analysis of covariance, are often used by social scientists to examine relationships between certain variables when other, potentially distorting variables are held statistically constant. Users of these methods sometimes do not realize that the partialled variable or covariate is assumed to be measured without error, and that if this is not the case, very misleading conclusions may be drawn.

Figure 3.9 provides a simple example. A correlation between latent variables A and B is assumed in the model to be due wholly to a third variable C that influences them both; removing the effect of this third variable should,

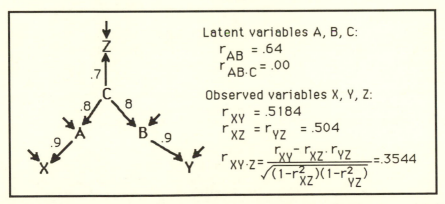

Fig. 3.9 Example of misleading partial correlation when partialled variable C is imperfectly measured by Z.

therefore, result in a partial correlation $r_{AB \cdot C}$ =0. But see what occurs if C happens to be measured with considerable error, and one applies the traditional formula for partial correlation. An observed correlation of .52 between A and B is reduced only to .35, perhaps leading an unsuspecting reader to believe that A and B are connected in other ways than through C. A structural analysis based on an appropriate path model, even with quite rough estimates of the reliabilities of measurement of A, B, and C, should provide a much less misleading picture.

Analyzing a quasi-experiment--the effect of Head Start

Table 3-11 presents some data from a study on the effects of a Head Start program on children's cognitive skills. Two measures of the latter were used--the Illinois Test of Psycholinguistic Abilities and the Metropolitan Readiness Test, both taken after completion of the Head Start program, as well as being given to a control group of nonparticipants.

As you can see from Table 3-11, the correlations between participating in Head Start and scores on the two tests assessing cognitive ability are on the face of it a little embarrassing to proponents of Head Start--although they are small, they are in the wrong direction: Participants in the program did a little worse than members of the control group.

Table 3-11 Correlations among variables in Head Start evaluation (data from Bentler & Woodward, 1978), N = 303

	MEd	FEd	FOc	Inc	HS	ITPA	MRT
Mother's education	1.00	.47	.24	.30	-.12	.26	.28
Father's education	.47	1.00	.28	.21	-.08	.25	.22
Father's occupation	.27	.24	1.00	.41	-.22	.22	.26
Income	.29	.23	.41	1.00	-.18	.12	.19
Head Start participation	-.10	-.11	-.21	-.18	1.00	-.10	-.09
ITPA score	.26	.23	.24	.15	-.10	1.00	.65
MRT score	.26	.23	.24	.15	-.10	.65	1.00

Note: ITPA = Illinois Test of Psycholinguistic Abilities, MRT = Metropolitan Readiness Test. Original correlations are above diagonal, correlations predicted by fitted model are below it.

But there were also negative correlations between Head Start participation and various parental educational and economic measures-- apparently the control group members were selected from families somewhat better off than those from which the Head Start children came. Could this account for the results?

Figure 3.10 represents a path model proposed by Bentler and Woodward (1978). (It is actually only one of several considered in their article, but we confine ourselves to this one.) The model involves five latent variables. The four main source variables include the independent variable, Head Start participation, and three variables describing family background--a general socioeconomic status variable (SES) common to all four of the observed socioeconomic indicators, and two variables capturing those specific aspects of education and economic circumstances that are independent of general SES. (As the diagram indicates, Bentler and Woodward assumed that these latter two might be correlated with each other, and that all three might be correlated with Head Start participation.)

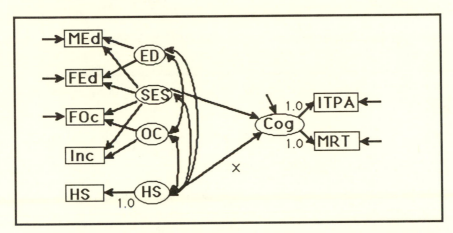

Fig. 3.10 Path diagram of a Head Start evaluation. x = path showing effect of Head Start participation on cognitive skills.

Other assumptions made in the diagram were that general SES and Head Start participation were the avenues of any influence of the source variables on cognitive skills, that cognitive skills were equally well measured by the two tests, and that Head Start participation and family income were measured without error. (One could certainly argue with some of these assumptions, but we proceed with the example. You might find it instructive to try fitting some other variations of the model to the data.)

The latent variables were taken as standardized except for *Cog*, which was assigned arbitrary paths of 1.0 to the observed variables (themselves standardized).

The crucial path, marked *x* in the diagram, describes the direct influence of Head Start on cognitive skills when the other variables are held constant. Is it positive (Head Start helps) or negative (Head Start hinders), and--in either case--does it differ significantly from zero?

97

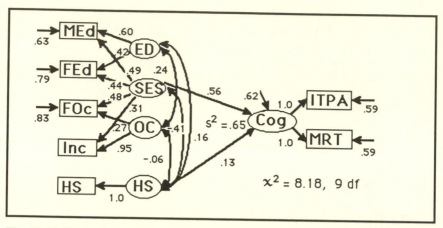

Fig. 3.11 Path diagram of Fig. 3.10 with values from solution including Head Start effect.

Figure 3.11 shows the values of the paths and the variance of the latent variable Cognitive Skills that were obtained by Bentler and Woodward in a solution using LISREL. Observe that when the socioeconomic statuses of the participants and the controls are taken into account, the estimate of the effect of Head Start participation on cognitive skills is slightly positive (.13).

The obtained χ^2 of 8.18 with 7 degrees of freedom represents a reasonably good fit to the data for the model as a whole, a fact that is also indicated by the reproduced correlations (shown below the diagonal in Table 3-11). However, Bentler and Woodward quite properly went on to test whether the particular path representing Head Start effects differed significantly from zero, by comparing the result with it included to that from a model that was identical except that this path was set to zero. The obtained χ^2 for the latter model was 9.88. The difference between the two χ^2s, 1.70, when tested as a χ^2 with 1 df, is well short of conventional levels of statistical significance (p > .10). Bentler and Woodward concluded, accordingly, that these data could not be said to demonstrate any effect of Head Start, positive or negative.

A note on degrees of freedom may be helpful. The model was treated as a covariance model with standardized latent independent variables. Twenty-eight (7 x 8 / 2) observed variances and covariances were fitted. Nineteen unknowns were solved for: 4 correlations among the source variables; 2 paths from source variables to *Cog* and 8 to the observed variables; the residual to *Cog* ; and 4 different residuals to observed variables (those to ITPA and MRT are forced by the model to be the same). Twenty-eight data points minus 19 unknowns equals 9 degrees of freedom. In the second model, with one less path to be solved for, there are 10 df.

98

Chapter 3 Notes

You might like to look at other published examples from the literature involving models like those in this chapter. Here is a sampling of some recent model-fitting studies on assorted topics (full citations are given in the reference list at the back of the book):

A. J. Fredericks & D. L. Dossett (1983)--a paper exploring attitude-behavior relationships.

J. E. Hunter, D. W. Gerbing, & F. J. Boster (1982)--Machiavellian beliefs and personality.

G. Donaldson (1983)--models of information-processing stages.

P. V. Marsden & K. E. Campbell (1984)--measuring the strength of ties between people.

W. M. Mason, J. S. House, & S. S. Martin (1985)--dimensions of political alienation.

M. D. Newcomb & P. M. Bentler (1983)--dimensions of the female orgasmic experience.

R. L. Baker, B. Mednick, & W. Brock (1984)--health problems in early infancy.

H. W. Marsh & D. Hocevar (1983)--a multitrait-multimethod analysis of university teaching evaluations.

H. W. Marsh & S. Butler (1984)--another MTMM paper, this time on tests diagnostic of reading problems.

Structural and measurement models. How they contribute to specifying the meaning of latent variables is discussed by R. S. Burt (1981).

Maruyama-McGarvey study. There is some inconsistency in the labeling of variables in the original paper. I have followed the identifications in their Table 2, which according to Maruyama (personal communication) are the correct ones.

Bramble-Wiley analysis. They actually used paths of $\pm 1/2$ rather than ± 1, which I have preferred for simplicity of presentation. This, of course, only affects the scale of the rawscore estimates, not the fit of the models or the standardized solution.

Multitrait-multimethod models. K. F. Widaman (1985) discusses hierarchically nested models for MTMM data, and Schmitt and Stults (1986) review different methods of analyzing MTMM matrices.

Chapter 3 Exercises

As before, other programs may be used for model-fitting, at the discretion of the instructor.

    1. Keep the measurement model from Maruyama and McGarvey's desegregation study but make one or more plausible changes in the structural model. Fit your model, using LISREL, and compare the results to those obtained by Maruyama and McGarvey.

    2. Solve McIver et al.'s police survey model (Fig. 3.4) using IPSOL, and compare to the results of Table 3-3. (Yes, it will require writing 36 equations.)

    3. Can you conclude that tests T1 to T3, whose covariance matrix is given below, are not parallel tests? (N = 35) How about *tau* -equivalent?

	T1	T2	T3
T1	54.85		
T2	60.21	99.24	
T3	48.42	67.00	63.81

    4. Part of Kelly and Fiske's original multitrait-multimethod matrix is given in Table 3-12. These are ratings of clinical psychology trainees by staff members, fellow trainees, and themselves. Tabulate and compare the correlations in the three principal categories (within trait, across method; within method, across trait; and across both method and trait).

    5. Estimate a multitrait-multimethod model for the data of Problem 4, using LISREL. (Assume methods are uncorrelated, and traits uncorrelated with methods.) Compare to the results of models using trait factors only and method factors only.

    6. Estimate a null model (all correlations zero) for the data of Problem 4. Calculate goodness-of-fit and parsimonious goodness-of-fit indices for the various solutions of Problem 5. Discuss your results.

    7. For the Bock et al. data (Table 3-9), use LISREL to test the hypothesis that the trait and acquiescence variances are equal. (Assume equal residual variances.)

Table 3-12 Multitrait-multimethod matrix (data from Campbell & Fiske, 1959), N = 124

| | As | Cs | Ss | Trait and method | | | | | |
				At	Ct	St	Ao	Co	So
Ratings:									
Staff									
Assertive	1.00								
Cheerful	.37	1.00							
Serious	-.24	-.14	1.00						
Trainee									
Assertive	.71	.35	-.18	1.00					
Cheerful	.39	.53	-.15	.37	1.00				
Serious	-.27	-.31	.43	-.15	-.19	1.00			
Self									
Assertive	.48	.31	-.22	.46	.36	-.15	1.00		
Cheerful	.17	.42	-.10	.09	.24	-.25	.23	1.00	
Serious	-.04	-.13	.22	-.04	-.11	.31	-.05	-.12	1.00

8. Calculate the original and partial correlations r_{XY} and $r_{XY \cdot Z}$--see Fig. 3.9--for the following additional values of path CZ: .9, 1.0, .5, .0. Comment on the results.

9. Construct a different path model for the Head Start evaluation data (Table 3-11), with different latent variables and hypothesized relations among them. Retain in your model the dependent latent variable of Cognitive skills and a path x to it from Head Start participation. Fit your model using LISREL and make a χ^2 test for the significance of path x.

Chapter Four:
Varieties of Path and Structural Models--II

In this chapter we continue our survey of a variety of applications of path and structural models. The models considered introduce some additional features over those discussed in Chapter 3. We begin by looking at a model containing reciprocal influences and correlated errors. Then we consider several models dealing with events over time. Finally, we consider models fitted across different groups.

Models With Reciprocal Influences and Correlated Errors

Most of the models considered so far have been unlooped, that is, they have no causal sequences that loop back on themselves either remotely (A causes B causes C causes A) or immediately (A causes B causes A). The latter are usually described as models with reciprocal influences, because A influences B and B influences A. Also, most of the models considered so far have uncorrelated residuals; that is, the miscellaneous unspecified residual causes that influence a given variable are assumed uncorrelated with any other specific or residual causes in the diagram.

Violations of either of these conditions create problems for Wright's rules and for path analyses carried out by ordinary regression methods, although it is possible to deal with them by various special techniques, such as two-stage least squares (e.g., James & Singh, 1978). However, these conditions do not present any special difficulties for the general iterative model-fitting procedures described in this book provided the models are adequately identified, that is, are sufficently well-rooted in data to yield definite solutions. In practice, such models often give trouble--assuring identification is sometimes not easy when loops or correlated errors are present, and convergence on a solution may be more difficult to attain--but in principle one simply specifies the model in the usual way and presents it to the iterative program. If the program can't solve the model, it responds in the same manner as it would for any other underspecified model or unsatisfactory set of starting values.

A study of career aspirations

Table 4-1 presents a set of data from a classic and much analyzed study by Duncan, Haller, and Portes (1968), in which both reciprocal influences and correlated residuals are involved. The original data were gathered in a study of career aspirations (Haller & Butterworth, 1960) in which 442 seventeen-year-old boys in a southern Michigan county were given tests and questionnaires. There were 329 boys in the sample naming at least one other boy also in the sample as one of their best friends. Thus, there were 329 instances of a boy and a close friend, on both of whom similar data on abilities, background, and career aspirations were available.

Table 4-1 Correlations among variables related to career aspirations of boys and their friends (data from Duncan, Haller, & Portes, 1968), N = 329

Respondent	Respondent					Friend				
	IQ	PA	SES	OA	EA	IQ	PA	SES	OA	EA
Intelligence	1.00	.18	.22	.41	.40	.34	.10	.19	.26	.29
Parent aspiration		1.00	.05	.21	.27	.08	.11	.02	.08	.11
Family SES			1.00	.32	.40	.23	.09	.27	.28	.31
Occup. aspiration				1.00	.62	.30	.08	.29	.42	.33
Educ. aspiration					1.00	.29	.07	.24	.33	.37
Friend										
Intelligence						1.00	.21	.30	.50	.52
Parent aspiration							1.00	-.04	.20	.28
Family SES								1.00	.36	.41
Occup. aspiration									1.00	.64
Educ. aspiration										1.00

Five variables were measured for each boy: (1) his intelligence (measured by his score on a nonverbal IQ test); (2) his perception of what his parents' aspirations were for his further education and occupational status; (3) his family's socioeconomic status (as measured by parental income and material possessions); (4) the level of occupation to which he aspired; and (5) the amount of further education that he expected to obtain.

Figure 4.1 shows a version of one model proposed by Duncan, Haller and Portes. In the left-hand part of the diagram are six source variables. Three represent the respondent's intelligence, his family's socioeconomic status, and his perception of his parents' aspirations for him; the other three represent the same measures for the respondent's friend. To the right in the diagram are two downstream latent variables representing the level of ambition of the respondent and his friend. It is assumed that an individual's ambition for

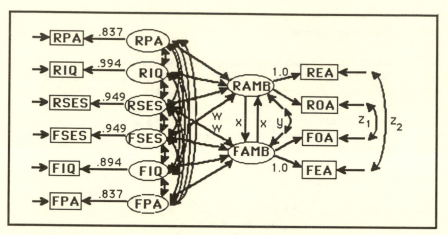

Fig. 4.1 Path model for study of career aspirations. R = respondent, F = friend; PA = parental aspirations; IQ = intelligence; SES = socioeconomic status; AMB = ambition; EA = educational aspirations;   OA = occupational aspirations. w, x, y, z = paths tested. Path values fixed by assumption are shown.

educational and economic success will be influenced by his parents' aspirations for him, by his intelligence, and by his family's socioeconomic status. It is also assumed (path marked *w*) that his friend's family's socioeconomic status might affect the level of education and occupation to which he himself aspires, but that his friend's intelligence and parental aspirations will not, except by way of the effect they might have on the friend's own ambition.

The latent variable of ambition is indexed by the two observed variables of educational aspiration and occupational aspiration, for both the respondent and his friend.

Several additional possibilities are considered, as shown on the path diagram:

1. Reciprocal influence, as indicated in Fig. 4.1 by the paths marked *x*. Does one's boy's ambition regarding educational and occupational achievement influence the ambition of his friend?

2. Correlated residual influences on the two friends' ambition, as indicated by the path marked *y*. It is possible that factors not included in the diagram that are shared in common by the two friends might influence their levels of ambition (in addition to whatever direct influence the boys have on each other). These might include, for example, the effect of teachers, or of other peers.

3. Correlated errors of measurement, as indicated by the paths marked *z*. There might be shared effects on the specifics of a particular measurement instrument, in addition to the true correlations of the common factors being

measured. For example, two friends might have discussed particular colleges and jobs, in addition to whatever general resemblance there was in their overall levels of ambition.

Previous analyses of these data have usually omitted the extreme left-hand part of the path diagram, i.e., the paths from the latent variables intelligence, SES, etc. to the observed measures of them. This is equivalent to assuming that these variables are perfectly measured--an unlikely assumption. If the reliabilities of the measurements were known, the square roots of these reliabilities could be entered as the values of the paths from the latent variables representing the true scores to the fallible observed scores (compare the example of Fig. 1.10, Chapter 1). For the purpose of illustration, we will arbitrarily assume that the reliabilities of measurement of parental aspiration, intelligence, and socioeconomic status are .7, .8, and .9, respectively, in this population, and hence that the square roots of these numbers, .837, .894, and .949, are the values of the corresponding paths in the measurement model.

A preliminary question we may ask is whether we need maintain a distinction between corresponding paths for respondents and their friends. After all, these are drawn from the same population, and it would not be at all surprising if they would agree within sampling error. If so, we need only solve for a single unknown value for each such set of paired paths, increasing the number of degrees of freedom and the general robustness of the analysis.

The first χ^2 test shown in Table 4-2 investigates this possibility. The first row of the table shows a χ^2 of 11.60, with 13 degrees of freedom, for the full model of Fig. 4.1. The second shows the results of assuming 15 equalities between respondent and friend: 3 in the correlations among the source variables for each individual (e.g., $r_{RPA,RIQ} = r_{FPA,FIQ}$); 3 in the correlations across pairs (e.g., $r_{RPA,FIQ} = r_{FPA,RIQ}$); 4 in the paths from source variables to Ambition (e.g., RPA to RAMB = FPA to FAMB); 1 in the reciprocal paths (RAMB to FAMB = FAMB to RAMB); 1 in the paths from the latent to the

Table 4-2. Tests of hypotheses for the Duncan-Haller-Portes career aspiration model (Fig. 4.1)

	χ^2	df	χ^2_{diff}	df	p
1. Unconstrained model	11.60	13			
2. Equality constraints only	18.89	28	7.29	15	>.90
3. No SES path (*w*)	23.07	29	4.18	1	<.05
4. No reciprocal influence (*x*)	20.81	29	1.92	1	>.10
5. No AMB residual correlation (*y*)	18.89	29	.00	1	>.95
6. No correlated errors (*z*)	32.21	30	13.32	2	<.01
7. Both 4 and 5 (no *x* or *y*)	25.67	30	6.78	2	<.05

Note: 2 tested against 1, and 3-7 against 2.

observed variables (RAMB to ROA = FAMB to FOA); and 3 in the residuals from occupational and educational aspirations and ambition.

The difference in χ^2 between the model with and without these equality constraints is 7.29; with 15 degrees of freedom this does not come remotely close to statistical significance. Thus, we might as well simplify matters by making the symmetry assumptions, which we do in the remainder of the analysis.

Line 3 asks whether we really need a path *w*. Is it necessary to postulate a direct influence of his friend's family's status on a boy's ambition? The result of the test is that the model fits significantly better with such a path than without it ($\chi^2 = 4.18$, 1 df, p < .05). However, the estimated value of the path (.08--see Table 4-3, model 2) suggests that it is not a major contributor to the determination of ambition. Line 4 asks the same question about reciprocal influences, the paths *x* in the figure. The model is judged not to fit significantly worse when they are omitted ($\chi^2 = 1.92$, 1 df, p > .10). The same conclusion can be drawn in line 5 about a possible correlation *y* between the residual factors lying back of the two latent dependent variables ($\chi^2 = .00$, 1 df, p > .95). Thus we cannot show that either of these features of the model--the influence of one friend's ambition on the other, or shared influences among the unmeasured variables affecting each--is necessary to explain the data. If, however, we exclude both of these simultaneously (the analysis of line 7) we *do* get a significant χ^2 (6.78, 2 df, p < .05), suggesting that the two may represent alternative ways of interpreting the similarity between friends' aspirations which our design is not sufficiently powerful to distinguish. As can be seen in Table 4-3, when both are fit, the residual correlation *y* is neglible (model 2), but when the reciprocal paths are set to zero (model 4), the correlation *y* becomes appreciable (.26). Setting *y* to zero (model 5) has little effect, as one would expect from its small previous value.

Finally, the analysis in line 6 of Table 4-2 asks about the possibility that the specific measures of educational and occupational aspirations might have errors that are correlated for the two friends. The substantial χ^2 (13.32, 2 df, p < .01) suggests that it is indeed a plausible assumption that such correlated errors exist.

This example, then, illustrates the application of a path diagram with somewhat more complex features than most of those we have considered previously. It is clear that further testable hypotheses could be stated for this model: For just one example, the diagram assumes that the respondent's own aspiration level and his estimate of his parents' aspirations for him are not subject to correlated errors. (Is this a reasonable assumption? How would you test it?) This case also suggests that tests of different hypotheses may not be independent of one another. In addition, if many hypotheses are tested, particularly if some are suggested by inspection of the data, one should remember that the nominal probability levels can no longer be taken literally,

Table 4-3 Estimated values of the paths and correlations for three models from Table 4-2

Paths	Model 2	Model 4	Model 5
PA to AMB	.19	.19	.19
IQ to AMB	.35	.38	.35
SES to AMB	.24	.26	.24
FSES to AMB (w)	.08	.12	.09
AMB to AMB (x)	.12	.00	.12
AMB to OA	.91	.91	.91
Correlations			
AMB residuals (y)	-.00	.26	.00
OA residuals (z_1)	.26	.26	.26
EA residuals (z_2)	.07	.06	.07

Note: Paths are unstandardized, but covariances have been standardized to correlations. Models correspond to the lines in Table 4-2.

though the differential χ^2s may still serve as a general guide to the relative merits of competing hypotheses.

Another point worth noting about this example is that *none* of the overall models tested in Table 4-2 can be rejected; that is, if one had begun with any one of them and tested only it, one would have concluded that it represented a tolerable fit to the data. It is only in the comparisons among the models that one begins to learn something of their relative merits.

Models of Events Over Time

As noted earlier, latent variable causal models are often used to analyze situations in which variables are measured over a period of time. Such situations have the advantage of permitting a fairly unambiguous direction of causal arrows: If event A precedes event B and there is a direct causal connection between them, it is A that causes B and not vice versa. If, on the other hand, A and B were measured more or less contemporaneously, a distinction between the hypotheses, A causes B, and B causes A, must be made on other grounds--not always a simple matter.

This is not to say that variables sequenced in time can never present difficulties in assigning cause. In the preceding example in which B follows A, it is always possible that B might be a delayed indicator of some third variable C that precedes and is a cause of A, and therefore one might be less wrong in calling B a cause of A than the reverse. Of course one would be still better off including C in the model as a cause of both A and B, with no causal arrow between A and B at all.

Nevertheless, there are many situations in which the presence of temporal ordering lends itself naturally to causal modeling, and we examine some examples in the next few sections.

A minitheory of love

Tesser and Paulhus (1976) carried out a study in which 202 college students filled out a 10-minute questionnaire on attitudes toward dating. The questionnaire contained several subscales having to do with attitudes and behavior toward a particular member of the opposite sex "where there is some romantic interest involved on somebody's part." Four measures were obtained: (T) how much the respondent thought about the other person during the last 2 weeks; (L) a 9-item love scale; (C) to what extent were the respondent's expectations concerning the other person confirmed by new information during the past 2 weeks; and (D) number of dates with the other person during the same 2-week period.

Two weeks later the subjects filled out the questionnaire again, with respect to the same person, for events during the 2 weeks between the two questionnaire administrations. Table 4-4 presents Tesser and Paulhus' basic results, which they subjected to a simple path analysis and which were later reanalyzed by Bentler and Huba (1979) using several different latent variable models.

Table 4-4 Correlations among four measures of "love" on two occasions (data from Tesser and Paulhus, 1976), N = 202

	T1	L1	C1	D1	T2	L2	C2	D2
Occasion 1								
Thought	1.000	.728	.129	.430	.741	.612	-.027	.464
Love		1.000	.224	.451	.748	.830	.094	.495
Confirmation			1.000	.086	.154	.279	.242	.104
Dating				1.000	.414	.404	.108	.806
Occasion 2								
Thought					1.000	.764	.161	.503
Love						1.000	.103	.505
Confirmation							1.000	.070
Dating								1.000
Mean	9.83	50.66	5.08	3.07	9.20	49.27	4.98	2.95
SD	3.59	19.49	1.80	2.87	3.75	20.67	1.72	3.16

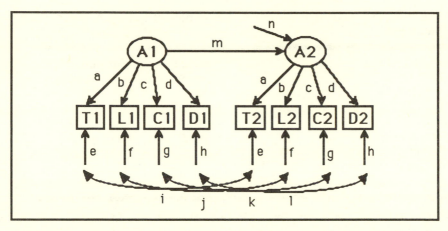

Fig. 4.2 A model for the "love" data of Tesser and Paulhus (Table4-4). A = general attraction; T, L, C, D = four measures of specific attitudes and behavior (see text); 1, 2 = two occasions.

Figure 4.2 shows a slightly modified version of one of Bentler and Huba's models. Basically, the four scales are shown as reflecting a common factor of attraction at each time period; attraction at the second period is explainable by a persistence of attraction from the first (path m) plus possible new events (path n). It is assumed that the measurement model (a, b, c, d ; e, f, g, h) is the same on both occasions of measurement. It is also assumed that the specifics of a particular behavior or attitude may show correlation across the two occasions. For example, an individual's frequency of dating a particular person is influenced by a variety of factors other than general attraction, and these might well be similar at both times--as might also be various measurement artifacts, such as the tendency of a person to define "dates" more or less broadly, or to brag when filling out questionnaires.

Table 4-5 shows the results of fitting the model of Fig. 4.2 to the correlations in Table 4-4. The paths reported are from an unstandardized solution (using LISREL); however, the measured variables are implicitly standardized by the use of correlations, the variance of the latent variable A1 is set to 1.0, and that of A2 does not differ much from 1.0, so the results in the table can pretty much be interpreted as though they were from a standardized path model. Thinking about a person and the love questionnaire are strong measures of the general attraction variable, dating is a moderate one, and confirmation of expectations is a very weak one. The residual factors, of course, reflect these inversely-- the love score is least affected by other things, and the confirmation score is nearly all due to other factors. The general factor of attraction toward a particular person shows a strong persistence over the 2 weeks (m = .94, standardized, .92). The residual covariances suggest that for thought and love the correlation between the two occasions of measurement is

109

Table 4-5 Solution of path model of Fig. 4.2 for data of Table 4-4: Tesser and Paulhus study

Variable	Paths		Residual paths		Residual covariances	
Thought	a	.83	e	.55	i	.11
Love	b	.88	f	.45	j	.09
Confirmation	c	.17	g	.98	k	.21
Dating	d	.53	h	.84	l	.53
Attraction	m	.94	n	.39		

Note: Paths unstandardized; variance of A1 set at 1.0, variance of A2 = 1.043. χ^2 = 45.87, 22 df, p < .01.

mostly determined by the persistence of the general factor, whereas for dating there is a large cross-occasion correlation produced by specific factors. On the whole, the measure of confirmation of expectations does not relate to much of anything else within occasions, and only quite moderately to itself across occasions. It was based on only one item; one might speculate that it might not be a very reliable measure. The measure of dating frequency may suffer from some psychometric problems as well--it appears to be markedly skewed (SD $\approx$ mean in Table 4-4). One might wish in such a case to consider preliminary transformation of the scale (say to logarithms or some such), before embarking on an analysis that assumes multivariate normality. Or at any rate, one should hedge on one's probability statements.

As a matter of fact, based on the obtained χ^2 of 45.87 with 22 degrees of freedom, if one takes the statistics seriously one would conclude that the present model does not fit the data (a conclusion that Bentler and Huba also arrived at from an analysis based on covariances using a similar model). If one calculates the correlations implied by the solution of Table 4-5 and compares them to the observed correlations, the largest discrepancies are for the correlation between T1 and C2, which the model predicts to be about .13 but which was observed as -.03, and for the correlation between C1 and L2, which was predicted as .14 but observed as .28. If one includes ad hoc paths for these in the model, the fit becomes statistically acceptable (χ^2 = 26.34, 20 df, p > .15)--Bentler and Huba obtained a similar result in their analysis. Because in doing this one is likely to be at least in part fitting the model to the idiosyncrasies of the present data set, the revised probability value should be taken even less seriously than the first. The prudent stance is that paths between T1 and C2 and C1 and L2 represent hypotheses that might be worth exploring in future studies but should not be regarded as established in this one.

Should one analyze correlations or covariances? As we have seen, in the present example, the results come out pretty much the same whether correlations were analyzed, as described, or whether covariances were, as in Bentler and Huba's analysis of these data. Both methods have their advantages. It is easier to see from the .83 and .88 in Table 4-5 that paths *a* and *b* are roughly comparable, than to make the same judgment from the values of 3.18 and 16.16 in Bentler and Huba's Table 1. On the other hand, the statistical theory underlying maximum likelihood and generalized least squares model fitting is based on covariance matrices, and application of these methods to correlation matrices, although widely practiced, means that the resulting χ^2s will contain one more step of approximation than they already do.

One further consideration, of minor concern in the present study, will sometimes prove decisive. If the variances of variables are changing markedly over time, one should be wary of analyzing correlations because this in effect restandardizes all variables at each time period. If one does not want to do this, but does wish to retain the advantages of standardization for comparing different variables, one should standardize the variables once, either for the initial period or across all time periods combined, and compute and analyze the covariance matrix of these standardized variables.

The simplex--growth over time

Suppose you have a variable on which growth tends to occur over time, such as height or vocabulary size among schoolchildren. You take measurements of this variable once a year, say, for a large sample of children. Then you can calculate a covariance or correlation matrix of these measurements across time: grade 1 versus grade 2, grade 1 versus grade 3, grade 2 versus grade 3, and so on.

In general, you might expect that measurements made closer together in time would be more highly correlated--that a person's relative standing on, say, vocabulary size would tend to be less different on measures taken in grades 4 and 5 than in grades 1 and 8. Such a tendency will result in a correlation matrix that has its highest values close to the principal diagonal and tapers off to its lowest values in the upper right and lower left corners. A matrix of this pattern is called a simplex (Guttman, 1954).

Table 4-6 provides illustrative data from a study by Bracht and Hopkins (1972). They obtained scores on standardized tests of academic achievement at each grade from 1 to 7. As you can see in the table, the correlations tend to show the simplex pattern by decreasing from the main diagonal toward the upper right-hand corner of the matrix. The correlations tend to decrease as one moves from left to right along any row, or upwards along any column. The standard deviations at the bottom of Table 4-6 show another feature often found with growth data: The variance increases over time.

Figure 4.3 represents a path diagram of a model fit by Werts, Linn, and

Table 4-6 Correlations and standard deviations across grades 1-7 for academic achievement (Bracht & Hopkins, 1972), Ns = 300 to 1240

Grade	Correlations						
	1	2	3	4	5	6	7
1	1.00	.73	.74	.72	.68	.68	.66
2		1.00	.86	.79	.78	.76	.74
3			1.00	.87	.86	.84	.81
4				1.00	.93	.91	.87
5					1.00	.93	.90
6						1.00	.94
7							1.00
SD	.51	.69	.89	1.01	1.20	1.26	1.38

Jöreskog (1977) to these data. Such a model represents one possible way of interpreting growth. It supposes that the achievement test score (T) at each grade level is a fallible measure of a latent variable, academic achievement (A). Achievement at any grade level is partly a function of achievement at the previous grade, via a path w, and partly determined by other factors, z. Test score partly reflects actual achievement, via path x, and partly random errors, u. Because variance is changing, it is appropriate to analyze a covariance rather than a correlation matrix. Covariances may be obtained by multiplying each correlation by the standard deviations of the two variables involved.

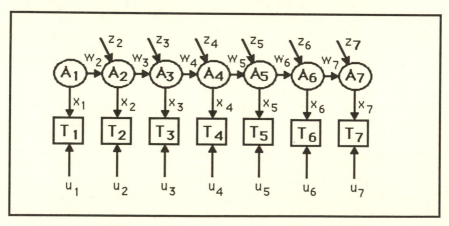

Fig. 4.3 Path model of growth over time. A = academic achievement; T = test score; 1-7 = grades.

Figure 4.3 has 7 *xs*, 7 *us*, 6 *ws*, and 6 *zs*, or a total of 26 unknowns. There are 7 x 8 / 2 = 28 variances and covariances to fit. However, as Werts et al. point out, not all 26 unknowns can be solved for: There is a dependency at each end of the chain so that two unknowns--e.g, two *us*--must be fixed by assumption. Also, they defined the scale of the latent variables by setting the *xs* to 1.0, reducing the number of unknowns to 18--5 *us*, 6 *ws*, and 7 *zs*--leaving 10 degrees of freedom. (Seven *zs* because A_1 is solved for as z_1 in the **PS** matrix.)

Table 4-7. Solution of path diagram of Fig. 4.3 for data of Table 4-6 (growth over time)

Grade	w	z	A	u
1			.184	.076[a]
2	1.398	.041	.400	.076
3	1.318	.049	.743	.049
4	1.054	.137	.962	.058
5	1.172	.051	1.372	.068
6	1.026	.104	1.548	.040
7	1.056	.138	1.864	.040[a]

Note: w,z,A,u as in Fig. 4.3. Values u[a] set equal to adjacent value of u. A,z,u expressed as variances, w as an unstandardized path coefficient. If N taken as 300, χ^2 = 10.76, 10 df, p > .30. $A_n = (T_n - u_n)/x^2_n$, where T_n is the variance of test n.

Table 4-7 shows estimates of the unknown values. The simplex model provides a reasonable fit to the data (χ^2 = 10.76, 10 df, p > .30), at least if N is taken equal to the lower end of its range. The variance of academic achievement, A, increases steadily and substantially over the grades, whereas trends for *w*, *z*, and *u* are much less marked, especially if one discounts the first 2 or 3 years.

A point of mild interest in this solution is that the *w* parameters, which represent the effect of academic achievement in one grade on that in the next, are slightly greater than 1.0. Does this mean that academic skill persists without loss from one year to the next, indeed with enhancement? Are students who think they forget things over the summer really mistaken? Alas, more likely it means that what happens to a student between one year's measurement and the next is correlated with his standing the preceding year, so that the academically rich get richer and the poor lag further behind them. A suitable latent variable analysis taking additional variables into account would provide a way to clarify this issue.

Finally, could we fit an even simpler model to these data, one that has *w*, *z,* and *u* constant, and only A varying? The answer can be obtained by fitting a model with just a single *z, w,* and *u.* The resulting χ^2 of 75.66 with 25 degrees of freedom says, No, we cannot. The grade-to-grade differences in these parameters are too large to be attributable merely to chance.

Liberal-conservative attitudes at three time periods

Judd and Milburn (1980) used a latent variable analysis to examine attitudes in a nationwide sample of individuals who were surveyed on three ocasions, in 1972, 1974, and 1976. Table 4-8 shows a portion of their data, based on three topics related to a liberal-conservative dimension of attitude (actually, Judd and Milburn studied five such topics). These particular data are from a subsample of 143 respondents who had attended 4 or more years of college. The numbers in the table mean, for example, that these respondents' attitudes toward busing in the 1972 and 1974 surveys were correlated .79, and their attitude toward busing in 1972 was correlated .39 with their attitude toward criminal rights in 1974.

The authors postulated that the interrelationships among these attitude measurements would largely be accounted for by a general factor of liberalism-conservatism, to which all three of the attitudes would be related at each of the

Table 4-8. Correlations among attitudes at three time periods (Judd & Milburn, 1980), N = 143, 4 years college

		B_{72}	C_{72}	J_{72}	B_{74}	C_{74}	J_{74}	B_{76}	C_{76}	J_{76}
1972	Busing	1.00	.43	.47	.79	.39	.50	.71	.27	.47
	Criminals		1.00	.29	.43	.54	.28	.37	.53	.29
	Jobs			1.00	.48	.38	.56	.49	.18	.49
1974	Busing				1.00	.46	.56	.78	.35	.48
	Criminals					1.00	.35	.44	.60	.32
	Jobs						1.00	.59	.20	.61
1976	Busing							1.00	.34	.53
	Criminals								1.00	.28
	Jobs									1.00
	SD	2.03	1.84	1.67	1.76	1.68	1.48	1.74	1.83	1.54

Note: Busing = bus to achieve school integration; criminals = protect legal rights of those accused of crimes; jobs = government should guarantee jobs and standard of living.

three time periods, plus a specific factor for each attitude that would persist across time. (Actually, the main focus of Judd and Milburn's interest was to compare these features of attitude in a relatively elite group, the present sample, with those in a non-elite group, consisting of respondents who had not attended college. We look at this aspect of the study in the next section of this chapter, which considers cross-group comparisons.)

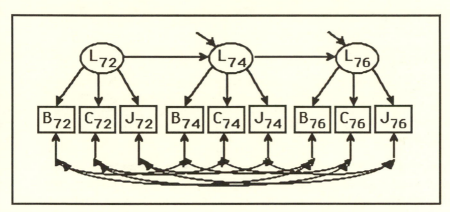

Fig. 4.4 Path model for attitudes measured in 1972, 1974, and 1976.   L = general, B,C,J = specific attitudes; 72,74,76 = years.

Figure 4.4 represents their hypothesis. Liberalism in 1974 is partly predictable from liberalism in 1972, and partly by unrelated events; and similarly for 1976. The general degree of a person's liberalism in any year is reflected in his or her specific attitudes toward busing, the rights of criminals, and guaranteed jobs. A person's attitudes on one of these specific topics in one survey is related to his or her attitude on this same topic in another survey, but not with specific attitudes on other subjects, except by way of the common liberalism-conservatism factor. (Actually, Judd and Milburn worked with a slightly different, but essentially equivalent, model.)

A word about setting up models of events over time in LISREL-type matrices may be helpful. Such models, like the present one, often have correlated errors for a given variable measured on different occasions. This means that such a variable cannot conveniently be represented on both the X and Y sides of the model because LISREL only provides matrices of correlations of residuals within the separate sets of X and Y variables. The model should, then, be set up only on one side--normally the Y side, to allow the paths connecting latent variables over time to be represented in the BE matrix. The PS matrix will then contain the residuals of the latent variables. (The source latent variable or variables at the initial time period can be regarded as having their total variance "residual" because it comes from causal sources outside those represented in the diagram.)

Table 4-9 Solution of path model of Fig. 4.4 representing liberal/conservative attitudes at three time periods

		Path from L	Residual variance	Specific covariance with 1974	Specific covariance with 1976
1972	Busing	1.00[a]	1.51	.58	.29
	Criminals	.58	2.52	.88	1.21
	Jobs	.63	1.74	.41	.37
1974	Busing	1.00[a]	.92		.23
	Criminals	.62	2.00		1.22
	Jobs	.68	1.18		.48
1976	Busing	1.00[a]	.72		
	Criminals	.49	2.82		
	Jobs	.61	1.49		

Note: Unstandardized coefficients. Paths marked[a] arbitrarily set at 1.00. χ^2 =11.65, 16 df, p > .70. Additional paths: L_{72} to L_{74} = .86, L_{74} to L_{76} = .99; L_{72} variance = 2.60; residual variances, L_{74} = .24, L_{76} = .18.

Table 4-9 presents an analysis of the Judd and Milburn data using LISREL and a covariance matrix based on Table 4-8.

On the whole, the model fits very well $(\chi^2$ = 11.65, 16 df, p > .70). Liberalism is most strongly defined by attitudes toward busing, with attitudes toward guaranteed jobs ranking slightly ahead of attitudes toward justice for accused criminals. Not surprisingly, the three attitudes tend to fall in the reverse order with respect to unexplained variance, as well as the amount of specific association with the same attitude in other years.

A question one might ask is whether liberal-conservative attitudes in 1972 would have any effect on those in 1976 except via 1974; i.e., could there be a delayed or sleeper effect of earlier on later attitudes? This can be tested by fitting a model with an additional direct path from L_{72} to L_{76}. This yields a χ^2 of 11.56 for 15 df. The difference, a χ^2 of .09 with 1 df, is far short of statistical significance. There is thus no evidence of such a sleeper effect in these data.

Models Comparing Different Groups

The general approaches described in this and the preceding chapter are readily extended to the case of model fitting in several independent groups of subjects. In the fitting process, one simply adds together the goodness-of-fit criterion from each of the separate groups and minimizes the total. For statistical tests, if a maximum likelihood or generalized least squares criterion is used, one derives an overall χ^2 which is the sum of the χ^2s in the separate groups, with an appropriate *df* which is the difference between the number of empirical values being fitted and the number of unknowns being solved for, taking into account any constraints being imposed within or across groups.

Again, differences in χ^2s for different nested solutions can be compared, using the differences between the associated degrees of freedom. Thus, for example, if one were solving for five unknowns in each of three groups, one could compare a solution that allowed them all to differ in each group with one that required four to be constant across groups and allowed only one to vary. There would be 15 unknowns to be solved for in the first case, and only $4 + 3 = 7$ in the second. So the increase in χ^2 between the two would be tested as a χ^2 with 15 - 7 = 8 df.

Attitudes in elite and non-elite groups

Earlier in this chapter we discussed a set of data by Judd and Milburn involving the structuring of attitudes with respect to a dimension of liberalism-conservatism. These attitudes were measured in three different years for a sample of 143 college-educated respondents. Responses were also available from the same nationwide surveys for a group of 203 individuals who had not attended college. (An intermediate group who had attended college, but for less than 4 years, were excluded in order to sharpen the contrast between the "elite" and "non-elite" groups.) Table 4-10 shows the data for the noncollege group, corresponding to Table 4-8 for the college group.

As we have seen, a model of a general attitude at each time period and specific attitudes correlated across time periods fit the data for college graduates quite well. Would it do as well for a less elite group? If it did, would there be quantitative differences between the groups in the various parameters of the model?

One can fit the model of Fig. 4.4 simultaneously to the data from both groups. If the same model fits in both, but with different values for the paths, one can conclude that the same general sort of explanation is applicable in both groups although with differences in quantitative detail. Or one can go further and ask if the same model with the same values will fit both sets of data. And, of course, one can take intermediate positions and constrain the values of certain paths to be the same in both groups, but allow others to vary.

If one fits the path model of Fig. 4.4 to the data of both the college and noncollege groups, without additional cross-group constraints, one obtains a χ^2 of 24.56 with 32 df, representing an excellent fit to the data (p > .80). (This in effect represents a separate solution for the same model in each group, and one can indeed do the solutions separately and add the χ^2s and dfs: fitting the model separately in the noncollege group gives a χ^2 of 12.91 with 16 df;

Table 4-10 Correlations among attitudes at three time periods (Judd & Milburn, 1980), N = 203, no college

		B_{72}	C_{72}	J_{72}	B_{74}	C_{74}	J_{74}	B_{76}	C_{76}	J_{76}
1972	Busing	1.00	.24	.39	.44	.20	.31	.54	.14	.30
	Criminals		1.00	.25	.22	.53	.21	.21	.40	.25
	Jobs			1.00	.22	.16	.52	.22	.13	.48
1974	Busing				1.00	.25	.30	.58	.13	.33
	Criminals					1.00	.21	.25	.44	.16
	Jobs						1.00	.21	.23	.41
1976	Busing							1.00	.17	.28
	Criminals								1.00	.14
	Jobs									1.00
	SD	1.25	2.11	1.90	1.31	1.97	1.82	1.34	2.00	1.79

Note: Busing = bus to achieve school integration; criminals = protect legal rights of those accused of crimes; jobs = government should guarantee jobs and standard of living.

11.65 + 12.91 = 24.56 and 16 + 16 = 32.)

If one goes to the opposite extreme and requires that both the model and quantitative values be the same in both groups, one obtains a χ^2 of 153.98 with 61 df, $p < .001$--thus, one can confidently reject the hypothesis of no quantitative differences between the samples.

One particular intermediate hypothesis, that quantities in the structural model are the same in both groups but the measurement models may be different, leads to a χ^2 of 27.05 with 37 degrees of freedom. This does not represent a significant worsening of fit from the original solution in which both structural and measurement models are allowed to vary ($\chi^2_{diff} = 2.49$, 5 df, $p > .70$). Thus, the difference between the two groups appears to lie in the measurement rather than the structural model.

Table 4-11 compares the solutions for the college and noncollege groups. The absolute values of paths from the latent variables to the observed variables are different for the two samples, but this is primarily a matter of the arbitrary scaling: attitude toward busing happens to be a relatively strong indicator of liberalism for the college group and a relatively weak one for the noncollege group, so that scalings based on this attitude will look quite different in the two cases. The standardized paths in the right-hand part of Table 4-11, obtained by

Table 4-11   Solution for the paths from liberalism to specific attitudes, for college and noncollege groups

		Unstandardized		Standardized	
		College	Noncollege	College	Noncollege
1972	Busing	1.00[a]	1.00[a]	.80	.63
	Criminals	.58	1.12	.51	.42
	Jobs	.63	1.40	.61	.58
1974	Busing	1.00[a]	1.00[a]	.84	.55
	Criminals	.62	.96	.54	.35
	Jobs	.68	1.44	.68	.58
1976	Busing	1.00[a]	1.00[a]	.87	.47
	Criminals	.49	.90	.41	.28
	Jobs	.61	1.65	.60	.58

Note: Paths marked[a] fixed at 1.0. Standard deviation for latent variable of liberalism from fitted solution: college--72 = 1.614, 74 = 1.475, 76 = 1.519; noncollege--72 = .789, 74 = .727, 76 = .633.

multiplying the unstandardized paths by the ratio of standard deviations of their tail to their head variables (see Chapter 1) provide a better comparison. Since the two samples are not very different in the overall level of variance of the observed variables (median SD across the 9 scales is 1.74 for college and 1.82 for noncollege), these values suggest a lesser relative contribution of the general liberalism-conservatism factor in the noncollege group.

Table 4-12 compares the paths between the latent variables across time. For both groups the analysis suggests a relatively high degree of persistence of liberal-conservative position, particularly between the 1974 and 1976 surveys.

Table 4-12   Solution for the paths connecting liberalism across years, for college and noncollege groups

	Unstandardized		Standardized	
	College	Noncollege	College	Noncollege
1972 to 1974	.86	.77	.94	.84
1974 to 1976	.99	.86	.96	.99

The genetics of numerical ability

Some problems in behavior genetics can be treated as straightforward intercorrelation or covariance problems involving multiple groups, and solved with programs such as LISREL, although more typically, explicit models are written and solved with general fitting programs. We will consider an example of each approach.

Table 4-13 gives correlations for three subscales of the Number factor in Thurstone's Primary Mental Abilities battery, in male and female identical and fraternal twin pairs. Correlations for male twins are shown above the diagonal in each matrix, and those for female twins are shown below. The data are from studies by S. G. Vandenberg and his colleagues in Ann Arbor, Michigan, and Louisville, Kentucky; the studies and samples are described briefly in Loehlin and Vandenberg (1968).

Table 4-13 Within-individual and cross-pair correlations for three subtests of numerical ability, in male and female identical and fraternal twin pairs (Ns: identicals 63, 59; fraternals 29, 46)

	Ad1	Mu1	3H1	Ad2	Mu2	3H2
Identical twins						
Addition 1	1.000	.670	.489	.598	.627	.456
Multiplication 1	.611	1.000	.555	.499	.697	.567
3-Higher 1	.754	.676	1.000	.526	.560	.725
Addition 2	.673	.464	.521	1.000	.784	.576
Multiplication 2	.622	.786	.635	.599	1.000	.540
3-Higher 2	.614	.636	.650	.574	.634	1.000
Fraternal twins						
Addition 1	1.000	.664	.673	.073	.194	.379
Multiplication 1	.779	1.000	.766	.313	.380	.361
3-Higher 1	.674	.679	1.000	.239	.347	.545
Addition 2	.462	.412	.500	1.000	.739	.645
Multiplication 2	.562	.537	.636	.620	1.000	.751
3-Higher 2	.392	.359	.565	.745	.603	1.000
	Ad1	Mu1	3H1	Ad2	Mu2	3H2
Standard deviations						
Identicals, male	7.37	13.81	16.93	8.17	13.33	17.56
Identicals, female	8.00	12.37	15.19	6.85	11.78	14.76
Fraternals, male	9.12	16.51	17.20	7.70	14.52	14.74
Fraternals, female	8.99	15.44	16.98	7.65	14.59	18.56

Note: In the correlation tables, males are shown above and females below the diagonal. 1 and 2 refer to scores of the first and second twin of a pair.

Figure 4.5 gives a path model for genetic influences on the correlations (or covariances) within and across twins. The latent variable N refers to a general genetic predisposition to do well on numerical tests. It is assumed to affect performance on all three tests, but perhaps to different degrees, as represented by paths *a, b, c.* These are assumed to be the same for both twins of a pair (designated 1 and 2). The genetic predispositions N are assumed to be perfectly correlated for identical twins, who have identical genotypes, but to be correlated .5 for fraternal twins, who are genetically ordinary siblings.

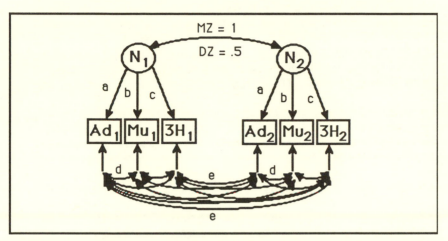

Fig. 4.5 Twin correlations on three subscales of numerical ability. MZ, DZ = identical and fraternal twins; N = genetic component of numerical ability; Ad, Mu, 3H = subscales; 1,2 =1st and 2nd twin of a pair.

The bottom part of Fig. 4.5 allows for nongenetic sources of correlation among abilities within individuals across pairs. Again, corresponding correlations are assumed to be equal--not all these are marked on the figure, but two examples are given. The residual correlation *d* between the addition and multiplication scales is assumed to be the same in those individuals designated "twin 2" as it is in those individuals designated "twin 1," and a correlation such as *e* between twin 1's score on "3-Higher" and twin 2's score on "Addition" is assumed to be the same as the correlation between twin 2's "3-Higher" score and twin 1's "Addition."

Altogether, there are 15 unknowns to be solved for: the 3 paths *a, b, c,* 3 residual variances, 3 within-person correlations across traits (*d* is an example), 3 different across-person correlations across traits (*e* is an example), and 3 across-person correlations for the same trait. There are $4 \times 6 \times 7 / 2 = 84$ data points (treating correlations as covariances), leaving $84 - 15 = 69$ df for testing the fit of the model to the data from the four groups at once. The obtained value of χ^2 (using LISREL) is 74.83, indicating that the model provides a reasonable explanation of the data (p > .20).

Could we improve things by fitting the model for the males and females separately? This would involve 30 unknowns and 84 - 30 = 54 df. The obtained χ^2 is 56.40, so the difference in χ^2 is 18.34 for 15 df, which does not represent a statistically significant improvement in fit (p >.20). We may as well go with the same result for both sexes.

Table 4-14 shows this result. The genetic paths have values from .58 to .73; the squares of these represent the proportion of variance attributable to genetic factors (if the model is correct), namely, from 33% to 54% for these three measures. The rest, the residual variances, are attributable to non-genetic factors, including errors of measurement. Whereas the trait variances include a component due to errors of measurement, the trait covariances do not. Here the genes show up more strongly, although the environmental contributions are still evident. The genetic contributions to the within-person

Table 4-14 Solution of model of Fig. 4.5 with data of Table 4-13, for genetics of numerical ability

	Genetic paths	Residual variances	Residual same-trait cross-person covariance
Addition	.680	.540	.138
Multiplication	.732	.468	.209
3-Higher	.575	.673	.378

	Other residual covariances	
	Within-person	Cross-person
Ad-Mu	.185	.092
Ad-3H	.244	.165
Mu-3H	.215	.204

correlations among the tests are .50, .39, and .42 (calculated from the path diagram as, for example, .680 x .732 = .50). The environmental contributions are .18, .24, and .22 (bottom of Table 4-14). The genes are estimated as contributing 63-74% to the correlations among the measures, as against 33-54% of their variance--reflecting the contribution of measurement error and specific environmental effects to the latter.

Heredity, environment, and sociability

In the last section we discussed fitting a model of genetic and environmental influences on numerical ability, treated as a LISREL problem involving multiple groups--namely, male and female identical and fraternal twins. In this section we consider a model-fitting problem in which data from two twin samples and a study of adoptive families are fit using a more general model-fitting program--in this case a variant of IPSOL.

The data to be used for illustration are given in Table 4-15. They are correlations on the scale "Sociable" of the Thurstone Temperament Schedule. The correlations are between pairs of individuals in the specified relationships. The first four pairings are for identical (MZ) and like-sexed fraternal (DZ) twins

Table 4-15. Correlations for the trait Sociable from the Thurstone Temperament Schedule in two twin studies and an adoption study

	Pairing	Correlation	Number of pairs
1.	MZ twins: Michigan	.47	45
2.	DZ twins: Michigan	.00	34
3.	MZ twins: Veterans	.45	102
4.	DZ twins: Veterans	.08	119
5.	Father-adopted child	.07	257
6.	Mother-adopted child	-.03	271
7.	Father-natural child	.22	56
8.	Mother-natural child	.13	54
9.	Adopted-natural child	-.05	48
10.	Two adopted children	-.21	80

Note: Michigan twin study described in Vandenberg (1962), and Veterans twin study in Rahe, Hervig and Rosenman (1978); correlations recomputed from original data. Adoption data from Loehlin, Willerman, and Horn (1985).

from two twin studies. The first study, done at the University of Michigan, involved highschool-age pairs, both males and females (see Vandenberg, 1962, for details). The second study was of adult pairs, all males, who had served in the U.S. armed forces during World War II and were located through Veterans Administration records (Rahe, Hervig, & Rosenman, 1978). The remaining pairings in the table are from a study of adoptive families in Texas (Loehlin, Willerman, & Horn, 1985).

Figure 4.6 shows a generalized path diagram of the causal paths that might underlie correlations such as those in Table 4-15. A trait S is measured in each of two individuals 1 and 2 by a test T. Correlation on the trait is presumed to be due to three independent sources: additive effects of the genes, G; nonadditive effects of the genes, D; and the environment common to pair members, C. A residual arrow allows for effects of the environment unique to each individual and--in all but the MZ pairs--for genetic differences as well.

Table 4-16 shows equations for the correlation $r_{T_1 T_2}$ between the test scores of members of various kinds of pairs. The equations are derived from

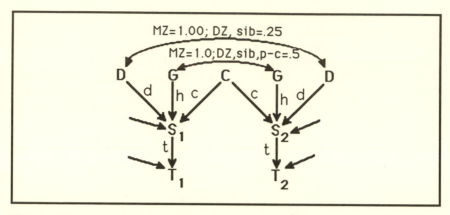

Fig. 4.6 Path model of genetic and environmental sources of correlation between two individuals. G = additive genes, D = nonadditive genetic effect, C = shared environment; S = sociability, T = test score; 1,2 = two individuals.

the path model of Fig. 4.6. The assumptions inherent in the genetic correlations at the top of Fig. 4.6 are that mating is random with respect to the trait; that all nonadditive genetic variance is due to genetic dominance; and that there is no selective placement for the trait in adoptions. Doubtless none of these is exactly true (for example, the spouse correlation in the adoptive familes for sociability was .16, which is significantly different from zero with 192 pairs but is certainly not very large). However, minor departures from the assumptions should not seriously compromise the model. The Table 4-16 equations allow (via c_1, c_2, c_3) for differentiating among the degrees of shared environment in the cases of identical twins, ordinary siblings, and parents and their children. The equations do not attempt to discriminate between the environmental relationships of parents and adopted or natural children, or of DZ twins and

Table 4-16 Equations for correlations between pairs of individuals in different relationships

Relationship	Table 4-15 pairings	Equation for correlation
MZ twins	1,3	$(h^2 + d^2 + c_1^2) \, t^2$
DZ twins	2,4	$(.5h^2 + .25d^2 + c_2^2) \, t^2$
Parent, adopted child	5,6	$(c_3^2) \, t^2$
Parent, natural child	7,8	$(.5h^2 + c_3^2) \, t^2$
Adoptive siblings	9,10	$(c_2^2) \, t^2$

Note: h, c, d, t as in Fig. 4.6.

other siblings; obviously, one might construct models that do, and even--with suitable data--solve them.

The path t in Fig. 4.6 is taken as the square root of the reliability of test T (the residual represents error variance). The reliability (Cronbach's alpha) of the TTS scale Sociable in the Veterans sample, .76, was assumed to hold for all samples. Thus, t was taken as $\sqrt{.76} = .87$ in solving the equations. A variant of IPSOL was used for the solution, modified to yield an approximate generalized least squares fit (for details, see Appendix B).

There are 10 observed correlations in Table 4-15; models with 1 to 4 unknowns were tested, allowing 6 to 9 df for the χ^2 tests.

Table 4-17 Solutions of Table 4-16 equations for various combinations of parameters

Model	χ^2	df	nfi	pfi	AIC
1. all rs equal (null)	36.02	9			-19.01
2. h^2 only	9.09	9	.75	.75	-5.54
3. $h^2 + c^2$	8.47	8	.76	.68	-6.24
4. $h^2 + d^2$	7.60	8	.79	.70	-5.80
5. $h^2 + c_1^2 + c_2^2$	6.18	7	.83	.64	-6.09
6. $h^2 + c_1^2 + c_2^2 + c_3^2$	2.60	6	.93	.62	-5.30

Table 4-17 gives χ^2s from several models based on the Table 4-16 equations. The first row contains a "null model"--that all correlations are equal. It can be rejected with confidence (p < .001). The models in the remaining lines of the table all constitute acceptable fits to the data (p > .30). We may still, however, compare them to see if some might be better than others. Adding a single environmental or nonadditive genetic parameter to h^2 (lines 3 or 4) does not yield a significant improvement in fit; nor does breaking down the environmental parameter into MZ twins versus others (line 5). A three-way breakdown of environment (line 6), into that for parent and child, siblings and MZ twins, does somewhat better, although the improvement is not statistically significant (χ^2_{diff} of 6.49 from line 2, 5.87 from line 3, and 3.58 from line 5, all p > .05).

The parsimonious fit index and Akaike's information criterion agree that solutions 3, 4, and 5 are inferior to the one-parameter simple genetic model of line 2, but they disagree on the four-parameter line 6 solution. The AIC evaluates it as the best of the lot, but the *pfi* judges it to be the poorest.

Table 4-18 gives parameter estimates for the two solutions that seem to deserve the most serious consideration, the one-parameter genetic model of line 2 and the four-parameter model of line 6.

The simple one-parameter genetic model estimates that approximately half of the variance of sociability (52%) is genetic in origin--the rest is presumably attributable to environment; this is not, however, the environment shared by family members, but that unique to the individual.

The model of line 6 allows for a more complex view of the effects of family environment, suggesting that the environmental effects on sociability are almost independent for parents and children, positively shared by identical twins, and negatively related for other siblings. If this last is correct, it suggests also that our model of a shared influence of environment, C in Fig. 4.6, is not altogether appropriate, at least in the case of siblings, because a variance should not be negative. Perhaps direct reciprocal negative paths between the individuals might provide a more satisfactory representation in this case. However, given that differentiating among the various environments does not produce a statistically significant improvement over the simpler model, such further elaborations are not pursued here.

Table 4-18 Parameter estimates for Table 4-17 solutions

---

line 2	$h^2 = .524$			
line 6	$h^2 = .449$	$c_{p-c}^2 = .022$	$c_{sib}^2 = -.167$	$c_{MZ}^2 = .153$

A Concluding Comment

The examples we have considered in this and the preceding chapter represent some typical applications of path and structural analysis to empirical data in the social and behavioral sciences. Most of these models happen to have been fit using LISREL, but as we noted earlier, this fact reflects the availability and ease of use of this particular program rather than any inherent feature of these kinds of problems.

In the next two chapters we turn temporarily away from models like these to consider the important class of latent variable methods known as exploratory factor analysis. In the final chapter we return to consider some further elaborations and refinements of path and structural models, and some other latent variable models sharing some of their features.

Chapter 4 Notes

Again, here is a sampling of recent papers involving some of the kinds of models discussed in this chapter:

G. W. Bohrnstedt & R. B. Felson (1983)--modeling self-esteem in the sixth to eighth grades.

D. C. Gottfredson (1982)--predicting who stays in school.

J. M. LaRocco (1983)--predicting who re-enlists in the Navy.

N. Schmitt (1982)--modeling different kinds of change in the attitudes of people who have lost jobs.

P. Lindsay & W. E. Knox (1984)--values related to work, at the time of highschool graduation and 7 years later.

E. S. Wellhofer (1984)--modeling political realignments in Great Britain between 1885 and 1950. Illustrates and compares to LISREL a different "soft modeling" approach called partial least squares--see, e.g., Wold (1982) and Dijkstra (1983).

R. R. Newton et al. (1984)--comparing the structure of anxiety in male and female college students.

S.-O. Brenner & R. Bartell (1984)--Is teacher stress the same in Canada and Sweden?

H. W. Marsh & D. Hocevar (1985)--Does the structure of the self- concept change across different grade levels?

D. Rindskopf & H. Everson (1984)--assessing racial discrimination in medical school admissions.

J. C. Loehlin (1985)--fitting heredity-environment models to data from a personality inventory in two twin samples and an adoption sample.

N. G. Martin, R. Jardine & L. J. Eaves (1984)--fitting genetic models to twin data on the National Merit Scholarship Qualifying Test.

Models with loops. Heise (1975) provides a good introduction to this topic, including the modifications of path rules required to deal with looped models.

Simplexes. Guttman (1954) originally proposed the simplex model for the case of a series of tests successively increasing in complexity, such that each required the skills of all the preceding tests, plus some new ones--an example would be addition, multiplication, long division. But simplex correlation patterns may occur in many other situations, such as the growth process considered in the chapter. Werts, Linn, and Jöreskog's example was slightly simplified for purposes of illustration by omitting data from one grade, the ninth. Jöreskog (Jöreskog & Sörbom, 1979, Chapter 3) discusses model-fitting involving a number of variants of the simplex. For a different approach to modeling growth, see the paper by McArdle cited in the next section.

Genetic assumptions. The value of .5 for the genetic correlation between dizygotic twins (numerical-ability example) assumes that assortative mating (the tendency for like to marry like) and genetic dominance and epistasis (nonadditive effects of the genes on the trait) are negligible in the case of numerical ability, or at least that to the extent they occur they offset one another. The first process would tend to raise the genetic correlation for fraternal twins, and the latter two would tend to lower it. Assortative mating tends to be substantial for general intelligence and verbal abilities but is usually modest for more specialized abilities, such as numerical and spatial skills (DeFries et al., 1979). In the sociability example, a path allowing for nonadditive genetic effects is included in the model.

For an ingenious method of using LISREL to fit genetic models to data from twins, see Boomsma and Molenaar (1986). McArdle (1986) discusses behavior genetic models of development and change. His models include means as well as covariances.

Chapter 4 Exercises

1. Repeat the test of the basic Duncan-Haller-Portes model of Fig. 4.1 (use the version with equality constraints--line 2 of Table 4-2). Then test to determine if each of the paths z_1 and z_2 makes a separate significant contribution to the goodness of fit of the model.

2. In the text, a question was raised about testing the assumption that a boy's aspiration for himself and his estimate of his parents' aspiration for him might be subject to correlated errors. How (in general) might this be tested? What problem would be presented if you were to attempt to test it using LISREL? How might you deal with this?

3. Fit the Tesser and Paulhus correlations (Table 4-4) as a confirmatory factor analysis involving five uncorrelated factors: a general attraction factor on which all eight measurements are loaded, and four specific factors, one for each test. Assume equal loadings across the two occasions for the general and the specific factors and the residuals.

4. For the model in Problem 3, relax the requirement that the loadings on the general factor are equal on the two occasions. Does this significantly improve the goodness of fit?

5. Repeat the analysis of the Judd and Milburn data (Tables 4-8 and 4-10), setting to 1.0 the paths to Jobs rather than the paths to Busing, as was done in the original analysis. Show the equivalence between the two solutions. (Note that covariances are analyzed, not correlations.)

6. Test the hypothesis for the Judd-Milburn data that the measurement model is the same across groups, although the structural model may differ.

7. Set up and solve the path problem for the genetics of numerical ability as in the text (sexes equal), for the correlations of Table 4-13. Still using correlations, test the additional hypothesis that the three subscales are parallel tests of numerical ability (i.e., have a single common parameter in each of the five sets in Table 4-14).

Chapter Five:
Exploratory Factor Analysis--I. Extracting the Factors

So far, we have been discussing cases in which a specific hypothesized model is fit to the data. Suppose we have a path diagram consisting of arrows from X and Y pointing to Z. The theory, represented in the path diagram, indicates that X and Y are independent causes, and the sole causes, of Z. The qualitative features of the situation are thus spelled out in advance, and the question we ask is, does this model remain plausible, once we look at the data? And if so, what are the quantitative features of the situation: What is our best estimate of the relative strengths of the two causal effects?

In this chapter we turn to another class of latent variable problems, the class that has been widely familiar to psychologists and other social and biological scientists under the name *factor analysis*, but which we are calling *exploratory factor analysis* to distinguish it from confirmatory factor analysis, which we have treated as an example of the kind of model fitting described in the preceding paragraph.

In exploratory factor analysis we do not begin with a specific model, only with rather general specifications about what kind of a model we are looking for. We must then find the model as well as estimate the values of its paths and correlations.

One can do a certain amount of exploration with general model-fitting methods, via trial-and-error modification of an existing model to improve its fit to data. But the methods we cover in this chapter and the next start out de novo to seek a model of a particular kind to fit to a set of data.

One thing that makes this feasible is that the class of acceptable models in the usual exploratory factor analysis is highly restricted: models with no causal links among the latent variables and with only a single layer of causal paths between latent and observed variables. (This implies, among other things, that these models have no looped or reciprocal paths.) Such models are, in the terminology of earlier chapters, mostly measurement model, with the structural model reduced to simple intercorrelations among the latent variables.

Indeed, in the perspective of earlier chapters, one way to think of

130

exploratory factor analysis is as a process of discovering and defining latent variables and a measurement model that can then provide the basis for a causal analysis of relations among the latent variables.

The latent variables in factor analysis models are traditionally called *factors*. Most often, in practice, both observed and latent variables are kept in standardized form; that is to say, correlations rather than covariances are analyzed, and the latent variables--the factors--are scaled to unit standard deviations. We mostly follow this procedure in this chapter. However, it is important to be aware that this is *not* a necessary feature of factor analysis--that one can, and in certain circumstances should, keep data in its raw score units and analyze covariances rather than correlations, and that some factor analytic methods scale factors to other metrics than standard deviations of 1.0.

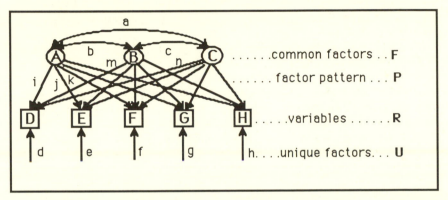

Fig. 5.1 Example of a factor analysis model. A, B, C = factors; D,E,F,G,H = observed variables; a,b,c = factor intercorrelations; d,e,f,etc. = specifics; i,j,k,etc. = factor pattern coefficients.

Figure 5.1 shows an example of a factor analysis model that reintroduces some of the factor analysis terminology that was earlier presented in Chapter 1 and adds a few new matrix symbols. A, B, and C are the three *common factors*. Their intercorrelations are represented by the curved arrows *a*, *b*, and *c*, which collectively form the *factor intercorrelation matrix*, which we designate **F**. D, E, F, G, and H are the *observed variables*, the tests or measures or other observations whose *intercorrelation matrix*, **R**, we are analyzing. The arrows *i*, *j*, *k*, etc. represent paths from latent to observed variables, the *factor pattern coefficients*. Collectively, these paths are known as the *factor pattern*, in matrix form **P**. Finally, paths *d*, *e*, *f*, etc. represent residual or *unique factors*, also called *specific factors*. They are expressed in matrix form as a diagonal matrix **U**, or as variances, U^2.

In the example, the dimensions of matrix **F** would be 3 x 3, matrices **R** and **U** would be 5 x 5 (although only the five nonzero diagonal values of **U** would be of interest), and **P** would be 5 x 3; conventionally, **P** is arranged so that the rows represent the observed variables and the columns the factors. Another matrix mentioned earlier, the *factor structure* matrix of correlations between factors and observed variables, is symbolized by **S**, its dimensions are also variables by factors, or 5 x 3 in the example. Recall that the elements of this matrix are a complex function of the paths and interfactor correlations--for example, the correlation between A and D is *i* + *bm* + *an*.

For a factor model, one can obtain the correlations implied by the model either by tracing the appropriate paths in the diagram according to Wright's rules, or, more compactly, by the matrix operations $_{imp}R = PFP' + U^2$, where the $_{imp}$ before **R** indicates that these are implied or predicted, rather than observed, values of the correlations. (**PFP'** by itself yields implied communalities in the diagonal instead of total variances--a so-called *reduced* correlation matrix that we symbolize by R_r.) The reader may wish to satisfy him or herself, by working through an example or two, that path tracing and matrix calculation indeed give identical results.

Now it is in general the case that there are an infinite number of possible path models that can reproduce any given set of intercorrelations, and this is still true even if we restrict ourself to the class of factor models. To give our search any point we must redefine it more narrowly. Let us invoke parsimony, then, and say that we are looking for the *simplest* factor model that will do a *reasonable* job of explaining the observed intercorrelations.

How does one determine whether a particular model does a reasonable job of explaining observed correlations? This is by now a familiar problem with a familiar solution: One generates the set of correlations implied by the model and then uses a formal or informal criterion of goodness of fit to assess their discrepancy from the observed correlations. Smallest absolute differences, least squares and maximum likelihood have all been used for this purpose.

What is meant by a *simple* model? Factor analysts typically use a two-step definition: (1) a model that requires the *smallest number of latent variables* (factors) ; (2) given this number of factors, the model with the *smallest number of nonzero paths* in its pattern matrix. Additional criteria are sometimes invoked, such as (3) uncorrelated factors, or (4) equal distribution of paths across variables or factors, but we focus on the first two, which are common to nearly all methods of exploratory factor analysis.

Applications of the first two criteria of simplicity correspond to the two main divisions of an exploratory factor analysis, *factor extraction* and *rotation.*

In the first step, factor extraction, methods are employed to yield models having the smallest number of factors that will do a reasonable job of explaining the correlations, although such methods typically produce models that are highly unsatisfactory according to the second criterion. Then in the second step, rotation, these models are transformed to retain the same small number of factors, but to improve them with respect to the second criterion of nonzero paths.

We consider methods of factor extraction in this chapter, and methods of rotation in the next.

Methods of Factor Extraction

One straightforward procedure goes as follows, beginning with the reduced correlation matrix R_r (which has estimated communalities replacing the 1s in the diagonal).

Step1. Solve for a general factor of R_r, using the centroid method described in Chapter 1, or other methods we discuss shortly.

Step 2. Obtain the matrix $_{imp}R_r$ implied by the obtained general factor and subtract it from the matrix used in Step 1, leaving a residual matrix that we designate $_{res}R$.

Step 3. Test the residual matrix $_{res}R$; are you willing to regard it as trivial? If so, stop. If not, put $_{res}R$ in place of R_r in Step 1, and repeat.

This account glosses over a few details, but it gives the essentials of a procedure that will produce a series of factors of decreasing magnitude, each of which is uncorrelated with all the others. This facilitates reaching the first goal of simplicity, the smallest number of factors necessary to fit the data reasonably well, because if factors are solved for in order of size, when one cuts off the process in Step 3, one knows that no potential factor remains unconsidered whose contribution toward explaining **R** would exceed that of the least important factor examined so far. And because the factors are independent, each obtained factor will make a unique and nonoverlapping contribution to the explanation of **R**.

The factors resulting from the process described, being general factors, will tend to have many nonzero paths and thus not be simple according to the second of our two criteria; but we deal with this problem later when we discuss the second stage of exploratory factor analysis known as "rotation."

Extracting a general factor

One method historically used for extracting a general factor, as we have noted, is the centroid method. Although it has largely been supplanted by the computer-based principal factor and maximum likelihood methods described shortly, the calculations of the centroid method are simple enough to be carried out readily by hand, so it is useful for illustrative purposes. An example is shown in Table 5-1. On the left in the table is an intercorrelation matrix, with communalities (in parentheses) replacing the 1's in the diagonal; thus, it is a reduced correlation matrix R_r. For purposes of the example, we have inserted exact communalities in the diagonal--ordinarily, of course, one would not know these, and would have to begin with estimates of them (we mentioned one possible method of communality estimation in Chapter 1 and discuss others later in this chapter).

Table 5-1 Extraction of an initial general factor by three methods (hypothetical correlations with exact communalities)

R_r						First general factor		
D	E	F	G	H		Centroid factor	Principal factor	Canonical factor
D (.16)	.20	.24	.00	.00	D	.205	.170	.065
E .20	(.74)	.58	.56	.21	E	.783	.782	.685
F .24	.58	(.55)	.41	.21	F	.680	.649	.525
G .00	.56	.41	(.91)	.51	G	.817	.857	.939
H .00	.21	.21	.51	(.36)	H	.441	.450	.507
.60	2.29	1.99	2.39	1.29				

sum = 8.56

√sum = 2.9257

The centroid calculations are carried through to arrive at the general factor shown in the first column on the right: i.e., the column sums of R_r are obtained, and summed, and divided by the square root of their sum, to yield the factor loadings given.

Also shown in Table 5-1 are general factors extracted from the same correlation matrix by two other methods. The column labeled *Principal factor* contains values obtained by an iterative search for a set of path coefficients which would yield the best fit of implied to observed correlations, according to a least squares criterion. The column labeled *Canonical factor* contains values obtained by a similar search using a maximum likelihood criterion instead. (The searches were carried out via LISREL, specifying one latent X variable, **LX** free, **PH** standardized, and **TD** fixed to U^2.)

Note that although each method leads to slightly different estimates of the paths from the factor to the variables, the solutions are generally similar, in that G is largest, D is smallest, with E then F and H falling between.

As we see later, there are other methods for obtaining principal and canonical factor loadings via the matrix attributes known as eigenvalues and eigenvectors, but those methods yield results equivalent to these.

Table 5-2 carries the process through to a second and third factor. The principal factor method is employed--either of the other two methods of obtaining general factors could be used in the same way. (If you try it with the centroid method, remember about the preliminary reflection of variables that is required when correlations are negative.)

In the first row of Table 5-2 are shown the correlation matrix, as in Table 5-1, with communalities in the diagonal. On the right, in the columns of factor pattern matrix **P**, the loadings of the three factors are entered as they are

Table 5-2 Extraction of three successive general factors by the principal factor method (data of Table 5-1)

R_r

	D	E	F	G	H
D	(.16)	.20	.24	.00	.00
E	.20	(.74)	.58	.56	.21
F	.24	.58	(.55)	.41	.21
G	.00	.56	.41	(.91)	.51
H	.00	.21	.21	.51	(.36)

P

	I	II	III	h^2
D	.170	.325	.161	.160
E	.782	.302	-.193	.740
F	.649	.330	.142	.550
G	.857	-.413	-.071	.910
H	.450	-.337	.208	.360

$_{res}R_1$

(.131)	.067	.130	-.146	-.076
.067	(.128)	.072	-.111	-.142
.130	.072	(.129)	-.146	-.082
-.146	-.111	-.146	(.175)	.124
-.076	-.142	-.082	.124	(.157)

$_{imp}R_1$

(.029)	.133	.110	.146	.076
.133	(.612)	.508	.671	.352
.110	.508	(.421)	.556	.292
.146	.671	.556	(.735)	.386
.076	.352	.292	.386	(.203)

$_{res}R_2$

(.025)	-.031	.023	-.012	.034
-.031	(.037)	-.028	.013	-.040
.023	-.028	(.020)	-.010	.029
-.012	.013	-.010	(.005)	-.015
.034	-.040	.029	-.015	(.043)

$_{imp}R_2$

(.106)	.098	.107	-.134	-.110
.098	(.091)	.100	-.124	-.102
.107	.100	(.109)	-.136	-.111
-.134	-.124	-.136	(.170)	.139
-.110	-.102	-.111	.139	(.114)

$_{res}R_3$

(-.001)	.000	.000	-.001	.001
.000	(.000)	-.001	-.001	.000
.000	-.100	(.000)	.000	-.001
-.001	-.001	.000	(.000)	.000
.001	.000	-.001	.000	(.000)

$_{imp}R_3$

(.026)	-.031	.023	-.011	.033
-.031	(.037)	-.027	.014	-.040
.023	-.027	(.020)	-.010	.030
-.011	.014	-.010	(.005)	-.015
.033	-.040	.030	-.015	(.043)

calculated. The first column, labeled I, is the first principal factor from Table 5-1, the single factor that by a least squares criterion comes closest to reproducing R_r. In the second row of matrices, on the right, are shown the correlations (and communalities) implied by the first general factor (obtained via **pp′**; e.g., $.170^2$ = .029; .170 x .782 = .133; etc.). On the left is what is left unexplained--the residual matrix $_{res}R$, obtained by subtracting $_{imp}R$ from R_r (e.g., .16-.029 = .131; .20 - .133 = .067; etc.)

The basic principal factor procedure is then applied to this residual matrix, to find the single general factor best capable of explaining these remaining correlations: The result is the second principal factor, labeled II in the matrix **P**. (This was again obtained by LISREL, with **TD** now fixed at 1-.131, etc.)

In the third row of matrices, these various steps are repeated. The matrix implied by factor II is $_{imp}R_2$, and the still unexplained correlations, $_{res}R_2$, are obtained by subtracting $_{imp}R_2$ from $_{res}R_1$. Clearly, not very much is left unexplained--the largest numbers in $_{res}R_2$ are on the order of .03 or .04. In many practical situations we might well decide that the small values left in $_{res}R_2$ are attributable to sampling or measurement error, less than perfect estimation of the communalities, or the like, and stop at this point. But in our hypothetical exact example we continue to a third factor, III, which, as shown in the bottom row, explains (except for minor rounding errors) everything that is left.

Note that the contributions of the three factors, that is, $_{imp}R_1$ + $_{imp}R_2$ + $_{imp}R_3$, plus the final residual matrix $_{res}R_3$, will always add up to the starting matrix R_r. This is a consequence of these being independent factors: Each explains a unique and nonoverlapping portion of the covariation in R_r. Note also that the sizes of the pattern coefficients tend on the whole to decrease as we move from I to II to III: Successive factors are less important; $_{imp}R_1$ explains more of R_r than does $_{imp}R_2$, and $_{imp}R_2$ more than $_{imp}R_3$.

Notice further that the total explained correlation R_r can be obtained either by $_{imp}R_1$ + $_{imp}R_2$ + $_{imp}R_3$ or by **PP′**. This equivalence is not surprising if one traces the steps of matrix multiplication, because exactly the same products are involved in both instances, and only the order of adding them up differs.

Finally, notice the column at the top right of Table 5-2 labeled h^2, the communalities implied by the solution. They are obtained as the diagonal of **PP′**, or, equivalently, as the sums of the squared elements of the rows of **P** (to see this equivalence, go mentally through the steps of the matrix multiplication **PP′**). In this case, because true communalities were used to begin with, and the solution is complete, the implied communalities agree with the diagonal of R_r.

Figure 5.2 compares the preceding solution, expressed in path diagram form, with the causal model which in fact was used to generate the correlation matrix analyzed in Table 5-2.

First, by the appropriate path tracing, either diagram yields the same correlations among variables and the same communalities. The communality of G in the top diagram is the sum of the squares of the paths to B and C, plus twice the product of these paths and the correlation r_{BC}; i.e., $.5^2+.6^2+2$x.5x.5x.6 = .91. The communality of G in the bottom diagram is just the sum of the squared paths to I, II, and III, because the latter are all uncorrelated; i.e., $.86^2+(-.41)^2+(-.07)^2 = .91$. The correlation between D and E in the top diagram

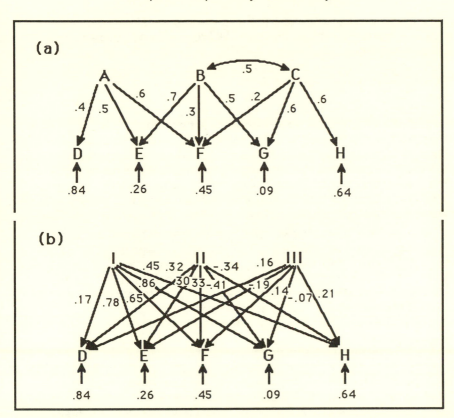

Fig. 5.2 Path models for Table 5-1. (a) Model used to generate correlations. (b) Model representing initial principal factor solution.

is .4x.5 = .20. That between D and E in the bottom diagram is .17x.78+.32x.30 +.16x(-.19), which also equals .20.

     Both these three-factor models, then, explain the data equally well: They imply the same correlations and communalities (and hence the same specific variances). The one explains the data with a smaller number of paths (9) and has two of its factors correlated. The other explains the data with three uncorrelated general factors of decreasing magnitude, involving a total of 15 paths, one from every factor to every variable.

     Most factor analysts believe that the action of causes in the real world is better represented by models like (a) than by models like (b). Causes typically have a limited range of effects--not every cause influences everything. And real-life causal influences may often be correlated in their actions. Nevertheless, a model like (b) has two great merits: (1) It can be arrived at by straightforward procedures from data, and (2) it establishes how many factors are necessary to explain the data to any desired degree of precision.

Furthermore, as was noted earlier, methods exist for transforming models like (b) into models more like (a), so that a model like (b) can be used as a first step in an exploratory analysis.

Direct calculation of principal factors

We have considered one way of arriving at an initial factor solution: by the successive extraction of independent general factors from a correlation matrix with communalities in the diagonal. In practice, however, a direct calculation can be used to obtain loadings for all the principal factors simultaneously. The availability of this calculation has contributed to making the principal factor method the most popular approach to exploratory factor analysis.

By this method, the principal factor pattern is obtained via the eigenvalues and eigenvectors of the reduced correlation matrix; i.e, the matrix $\mathbf{R}_r$. (Readers unfamiliar with the concepts of eigenvalues and eigenvectors should consult Appendix A or a matrix algebra text.) If we arrange the eigenvectors in the columns of a matrix $\mathbf{V}$ and the square roots of the eigenvalues from large to small in a diagonal matrix $\mathbf{L}$, we can obtain the principal factors by the matrix multiplication $\mathbf{P} = \mathbf{VL}$. Put another way, the principal factor pattern is a rescaling of the eigenvectors by the square roots of the eigenvalues. Postmultiplying a matrix by a diagonal matrix rescales its columns by the values in the diagonal matrix.

Given the eigenvalues and eigenvectors, then, the principal factor solution is simple. This just sweeps the computational effort under the rug, of course, by pushing it back into the chore of computing the eigenvalues and vectors. This is a very substantial computation, if carried out by hand for a large correlation matrix--think in terms of days or weeks, not minutes or hours. But fast and efficient computer routines exist for calculating the eigenvalues and vectors of symmetric matrices. Such routines are widely available, and access to one is assumed henceforward in this book; if for any reason you wish to solve for eigenvalues and vectors by hand, which is feasible for small examples, standard textbooks (e.g., Morrison, 1976) will show you how.

The eigenvalues corresponding to the principal factors are of interest in their own right--they represent the variance of observed variables explained by the successive factors. If we sum the squares of the factor loadings $\mathbf{P}$ in Table 5-2 by columns rather than rows, we will obtain the eigenvalues. They are, respectively, 2.00, .59, and .13. Their sum, 2.72, is the same as the sum of the communalities; it is the total explained variance. The first factor accounts for a substantial part of the total communality (2.00/2.72 of it, or about 74%). The second factor accounts for about 22%, and the third for 5%. Another way of looking at the three eigenvalues is as the sums of the diagonal elements (traces) of the three implied matrices in Table 5.2. (Can you see that these are algebraically equivalent?) Again, the eigenvalues reflect the relative contributions of the three factors.

Rescalings--Alpha and Canonical factors

We saw earlier that general factors may be obtained using a maximum likelihood rather than a least squares criterion, and we called these Canonical factors. It turns out that Canonical factors (Rao, 1955; Harris, 1962), and yet another variety, Alpha factors (Kaiser & Caffrey, 1965) can be obtained via the same basic eigenvalue-eigenvector procedure as principal factors, by rescaling the starting correlation or covariance matrix before carrying out the calculation.
    The two methods use different preliminary scalings: Alpha analysis rescales variables so that they have equal communalities of 1.0, whereas the Canonical factor approach rescales variables so that they have equal uniquenesses of 1.0.
    A numerical example of Alpha and Canonical factor analysis, based on the correlation matrix of Table 5-1, is given in Appendix D. The results are shown in Table 5-3. A general similarity to the centroid and principal factor solutions of Table 5-2 is evident, although there are differences in the sizes of coefficients. Those for the principal factor method tend to lie between the values derived from the two rescalings.

Table 5-3 Alpha and Canonical factor pattern matrices (correlation matrix of Table 5-1, exact communalities)

Alpha factors (rescaled)			Canonical factors (rescaled)		
I	II	III	I	II	III
.23	.31	.10	.06	.35	.18
.79	.15	-.31	.69	.51	-.12
.70	.23	.03	.53	.48	.20
.77	-.55	-.11	.94	-.17	-.01
.42	-.38	.20	.51	-.22	.24

Rao arrived at his formulation of Canonical factor analysis via an attempt to define factors that would have maximum generalizability to other samples of subjects. Kaiser and Caffrey arrived at their formulation of Alpha factor analysis via an attempt to define factors that would have maximum generalizability to other measures of the underlying variables. Although it is not necessarily the case that transformed versions of these solutions would retain these properties for individual factors, one might perhaps still choose one of these two approaches if one's primary concern lay with the one or the other kind of generalization.
    Both the Alpha and Canonical factor methods can be said to be "scale free," in the sense that they yield the same factors when starting from differently scaled variables: for example, from a covariance matrix of variables in their

original raw units, or from a correlation matrix, where the variables are implicitly standardized. The principal factor approach will give different factor solutions in these two cases. The Alpha and Canonical approaches, because of their preliminary rescaling of both the correlation and covariance matrices to the same standard form, will not, arriving at the same solution in each case. These factors--as in Table 5-3--are often scaled back to an ordinary standard-score metric at the end for interpretability. However, the basic properties of the solutions--maximum accounting for communality by each factor, and so on-- apply, of course, to the scaling in which the eigenvalue solution is actually carried out.

From a more general perspective, we may speak of various possible alternative scalings of variables for a factor analysis: (1) Leave the variables in their original rawscore metrics, i.e, do a principal factor analysis of the covariance matrix C (actually, of the reduced covariance matrix C_r, with common variances in the diagonal); (2) scale the variables by the square roots of their variances (their standard deviations), by factoring R_r; (3) scale the variables by the square roots of the common portions of their variances, i.e., do an Alpha analysis; or, (4) scale the variables by the square roots of the unique portions of their variances, i.e., do a Canonical analysis.

Alternative (2), the factor analysis of ordinary correlations, is by far the most widely used in practice. It might perhaps be regarded as a compromise between (3) and (4) when one is concerned, as one usually is, with generalizing across both subjects and variables. Alternative (1), the factoring of covariances, may suffer excessively from arbitrariness of scale: A variable, e.g., annual income, can have a quite different effect on the factor analysis if it is expressed in dollars or in thousands of dollars, because of the huge difference in the size of the variance and covariances in the two cases. However, when differing sizes of variances are at the heart of the issue, as may be the case in comparing factor analyses across different groups (e.g., different cultures, different ages, or the like), one would not want to lose the differences among the groups by restandardizing for each, and hence would prefer to work with the covariance matrices directly. As we have noted earlier, a possible way to eat one's cake and have it too is to standardize all one's variables over the combined groups, to deal with the problem of noncomparable units of the different variables, and then to factor analyze covariance matrices for the separate groups using this common metric.

We need now to return to two matters that we have so far finessed in our examples: namely, (1) estimating the communalities, and (2) deciding at what point the residuals become negligible. In real-life data analyses we do not usually have advance knowledge of how much of a variable's variance is shared with other variables and how much is specific. And in real life, we will usually have many trivial influences on our variables in addition to the main causes we hope to isolate, so that after the factors representing the latter are extracted we still expect to find a certain amount of residual covariance. At what point do we conclude that all the major factors have been accounted for, and what is left in the residual matrix is just miscellaneous debris? We consider these topics in turn.

Estimating Communalities

As we have seen, an exploratory factor analysis begins by removing unique variance from the diagonal of the correlation or covariance matrix among the variables. Because one rarely knows in advance what proportion of the variance is unique and what is shared with other variables in the matrix (if one did, one would probably not need to be doing an exploratory analysis), some sort of estimate must be used. How does one arrive at such an estimate? How important is it that the estimate be an accurate one?

The answer to the second question is easy: The larger the number of variables being analyzed, the less important it is to have accurate estimates of the communalities. Why? Because the larger the matrix, the less of it lies in the diagonal. In a 2 x 2 matrix, half the elements are diagonal elements. In a 10 x 10 matrix, only one-tenth are (10 diagonal cells out of a total of 100). In a 100 x 100 matrix, 1% of the matrix is in the diagonal, and 99% consists of off-diagonal cells. In a 2 x 2 matrix, an error in a communality would be an error in one of two cells making up a row or column total. In a 100 x 100 matrix, it would be an error in one of a hundred numbers entering into the total, and its effect would be greatly attenuated. In factoring a correlation matrix of more than, say, 40 variables, it hardly matters what numbers one puts into the principal diagonal, even 1's or 0's--although since it is very easy to arrive at better estimates than these, one might as well do so. Many different methods have been proposed. We have mentioned one earlier (the method of triads--see Chapter 1). We discuss two more in this chapter, plus a strategy for improving any initial estimate via iteration.

Highest correlation of a variable

A very simpleminded but serviceable approach is to use as the communality estimate for a given variable the highest absolute value of its correlation with any other variable in the matrix; that is, the largest off-diagonal number in each row in the matrix is put into the diagonal with positive sign.

The highest correlation of a variable with another variable in the matrix *isn't* its communality, of course, but it will in a general way resemble it: Variables that share much variance with other variables in the matrix will have high correlations with those variables and hence get high communality estimates, as they should, whereas variables that don't have much in common with any other variables in the matrix will have low correlations and hence get low communality estimates, again correctly. Some cases won't work out quite so well--e.g., a variable that has moderate correlations with each of several quite different variables might have a high true communality but would receive only a moderate estimate by this method. Nevertheless, in reasonably large matrices, or as a starting point for an iterative solution, this quick and easy method is often quite adequate.

Squared multiple correlations

A more sophisticated method, but one requiring considerably more computation, is to estimate the communality of a given variable by the squared multiple correlation of that variable with all the remaining variables in the matrix. In practice, this is usually done by obtaining R^{-1}, the inverse of the (unreduced) correlation matrix R. The reciprocals of the diagonal elements of R^{-1}, subtracted from 1, yield the desired squared multiple correlations (often called SMCs for short); that is, for the i th variable:

$$SMC_i = 1 - 1/k_{ii},$$

where k_{ii} is the i th element of the main diagonal of R^{-1}.

Table 5-4 illustrates the calculation of SMCs for the example of Table 5-1.

Table 5-4 Calculation of squared multiple correlations of each variable with all others (data of Table 5-1)

R	1.00	.20	.24	.00	.00
	.20	1.00	.58	.56	.21
	.24	.58	1.00	.41	.21
	.00	.56	.41	1.00	.51
	.00	.21	.21	.51	1.00

R^{-1}	1.096	-.204	-.230	.219	-.021
	-.204	1.921	-.751	-.869	.197
	-.230	-.751	1.585	-.189	-.079
	.219	-.869	-.189	1.961	-.778
	-.021	.197	-.079	-.778	1.372

	diagonal	1/diag.	SMC
D	1.096	.912	.088
E	1.921	.521	.479
F	1.585	.631	.369
G	1.961	.510	.490
H	1.372	.729	.271

R is the correlation matrix; **R**$^{-1}$ is its inverse, calculated by a standard computer routine. The bottom part of the table shows the steps in obtaining the SMCs.

SMCs are not communalities either; in fact, they are systematically lower than (at most equal to) the true communalities. Nevertheless, they are related to the communalities in a general way, in that if a variable is highly predictable from other variables in the matrix, it will tend to share a good deal of variance in common with them, and if it is unpredictable from the other variables, it means that it has little common variance. In large matrices, the SMCs are often only slightly below the theoretical true communalities.

Iterative improvement of the estimate

The basic idea, mentioned briefly in Chapter 1, is that one makes an initial communality estimate somehow, obtains a factor pattern matrix **P**, and then uses that to obtain the set of communalities implied by the factor solution. In the usual case of uncorrelated initial factors, these are just the sums of the squares of the elements in the rows of **P**; more generally, they may be obtained as the diagonal of **PFP′**, where **F** is the matrix of factor intercorrelations.

One can then take these implied communalities, which should represent a better estimate than the initial ones, put them in place of the original estimates in **R**$_r$, and repeat the process. The **P** from this should yield still better estimates of the communalities, which can be reinserted in **R**$_r$, and the process repeated until successive repetitions no longer lead to material changes in the estimates. Such a process involves a good deal of calculation, but it is easily programmed for a computer, and most factor analysis programs provide iterative improvement of initial communality estimates as an option.

Table 5-5 shows several different communality estimates based on the artificial example of Table 5-1. The first estimate, highest correlation in the row,

Table 5-5 Comparison of some communality estimates for the correlation matrix of Table 5-1

Variable	h^2	h^2 estimate 1	2	3	4
C	.16	.24	.09	.19	.13
D	.74	.58	.48	.73	.69
E	.55	.58	.37	.52	.61
F	.91	.56	.49	.81	.96
G	.36	.51	.27	.46	.34
	2.72	2.47	1.70	2.71	2.73

Note: h^2 = communality from model which generated r's. Estimates: (1) highest r in row; (2) SMC; (3) SMC with iteration (3 principal factors); (4) SMC with iteration (3 Alpha factors).

shows a not-atypical pattern for this method of overestimating low communalities and underestimating high ones. The second, SMCs, shows, as expected, all estimates on the low side. Columns 3 and 4 show the outcome of iterative solutions starting with SMCs: Column 3 shows a solution based on a principal factor analysis, and column 4 shows a solution based on an Alpha factor analysis. No solution recovers the exact set of communalities of the model generating the correlations, but the iterative solutions come much closer than either of the direct estimates, and the total estimated communality is also fairly close to that of the theoretical factors.

One disadvantage of iterative solutions for the communalities is that they will sometimes lead to a "Heywood case"; a communality will converge on a value greater than 1.0. This is embarrassing; a hypothetical variable that shares more than all of its variance with other variables is not too meaningful. Some factor analysis computer programs will stop the iterative process automatically when an offending communality reaches 1.0, but this really isn't much better, because a variable with no unique variance is usually not plausible either. A possible alternative strategy in such a case might be to show, e.g., by means of a χ^2 test, that the fit of the model with the communality reduced to a sensible value is not significantly worse than it is with the Heywood case communality. If this proves *not* to be the case, the model is unsatisfactory and something else must be considered--extracting a different number of factors, rescaling variables to linearize relationships, eliminating the offending variable or one of its correlates, or the like.

If you are a very alert reader, it may have occurred to you that there is another potential fly in the ointment in using iterative approaches. In order to use such an approach to improving communality estimates, one must first know how many factors to extract--because using a different number of columns in **P** will result in different implied communalities. In the case of our hypothetical example, we used the three factors known to account for the data as the basis of our iterative improvement, but in real life one must first decide how many factors to use if one wishes to iteratively improve communality estimates. To this problem of determining the number of factors we now turn.

Determining the Number of Factors

In practice, deciding on the number of factors is a much stickier problem than communality estimation. As mentioned in the last section, with reasonably large correlation matrices even quite gross errors in estimating the communalities of individual variables will usually have only minor effects on the outcome of a factor analysis. Not so with extracting too many or too few factors. This will not make too much difference in the initial step of factor extraction, other than adding or subtracting a few columns of relatively small factors from the factor pattern matrix **P**. But it will often make a material difference when the next, transformation stage is reached. Admitting an additional latent variable or two often leads to a substantial rearrangement of paths from existing latent variables; trying to fit the data with one or two fewer latent variables can also lead to a substantial reshuffling of paths. Such rearrangements can lead to quite different interpretations of the causal structure underlying the observed correlations.

So the problem is not a trivial one. What is its solution? As a matter of fact, several possible solutions have been proposed. We describe four: the Kaiser-Guttman rule, the scree test, a chi-square test for further significance of the residuals, and a cross-validation procedure due to Cudeck and Browne.

The Kaiser-Guttman rule

This is easily stated: (1) Obtain the eigenvalues of the correlation matrix **R** (*not* the reduced matrix **R**$_r$); (2) ascertain how many eigenvalues are greater than 1.0. That number is the number of nontrivial factors that there will be in the factor analysis.

Although various rationales have been offered for the choice of the particular value 1.0, none is entirely compelling, and it is perhaps best thought of as an empirical rule that often works quite well. Because it is easy to apply and has been incorporated into various popular computer programs for factor analysis, it has undoubtedly been the method most often used to answer the question "how many factors?" in factor analyses conducted over the last decade or two.

It is not, however, infallible. If you apply it, for example, to a set of eigenvalues obtained by factoring the intercorrelations of random data, the Kaiser-Guttman rule will not tell you that there are no interpretable factors to be found. On the contrary, there will typically be a sizeable number of factors from such data with eigenvalues greater than 1.0, so the rule will tell you to extract that many factors. (To see that there must be eigenvalues greater than 1.0, consider that their sum must be m for an m-variable matrix. When you extract them in order of size, there will be some larger than 1.0 at the beginning of the list and some smaller than 1.0 at the end.)

Table 5-6 provides an example, in which eigenvalues from the correlations of random scores and real psychological test data are compared. If one were to apply the Kaiser-Guttman rule to the random data, it would suggest the presence of 11 meaningful factors; there are, of course, actually none. For the real psychological data, the rule would suggest 5 factors, which is not unreasonable--factor analysts, using various criteria, have usually argued for either 4 or 5 factors in these particular data. (Note that the 5th eigenvalue is only just slightly above 1.0, which suggests another difficulty with a Kaiser-Guttman type of rule: Chance fluctuations in correlations might easily shift a borderline eigenvalue from, say, .999 to 1.001, leading to a different decision for the number of factors, but would one really want to take such a small difference seriously?)

Table 5-6. Eigenvalues from random and real data

Rank in size	Random data	Real data	Rank in size	Random data	Real data
1	1.737	8.135	13	.902	.533
2	1.670	2.096	14	.850	.509
3	1.621	1.693	15	.806	.477
4	1.522	1.502	16	.730	.390
5	1.450	1.025	17	.717	.382
6	1.393	.943	18	.707	.340
7	1.293	.901	19	.672	.334
8	1.156	.816	20	.614	.316
9	1.138	.790	21	.581	.297
10	1.063	.707	22	.545	.268
11	1.014	.639	23	.445	.190
12	.964	.543	24	.412	.172

Note: Random data = correlation matrix of random scores on 24 variables for 145 cases. Real data = Holzinger-Swineford data on 24 ability tests for 145 7th- and 8th-grade children, from Harman (1976, p. 161).

Presumably, one does not often factor correlations based on random data intentionally, but one may occasionally want to factor analyze something similar--say, intercorrelations of measures of quite low reliability, such as individual questionnaire items, which could be substantially influenced by random measurement errors. In such cases one could be led badly astray by blind reliance on the Kaiser-Guttman rule.

The scree test

This procedure also employs eigenvalues. However, instead of using a 1.0 cutoff, the user plots successive eigenvalues on a graph and arrives at a decision based on the point at which the curve of decreasing eigenvalues changes from a rapid, decelerating decline to a flat gradual slope.

The nature of this change can be best illustrated by an example. The eigenvalues for the real data from Table 5-6 are plotted in Fig. 5.3. Notice how the first few eigenvalues drop precipitously, and then after the fourth, how a gradual linear decline sets in. This decline is seldom absolutely linear out to the last eigenvalue--often, as here, it may shift to a more gradual slope somewhere enroute. This linear or near-linear slope of gradually declining eigenvalues was called the *scree* by R.B. Cattell (1966a), who proposed this test. He arrived at this name from the geological term for the rubble of boulders and debris extending out from the base of a steep mountain slope. The idea is that when you come up to the top of the scree, you have reached the real mountain slope--or the real factors. Below that, you have a rubble of trivial or

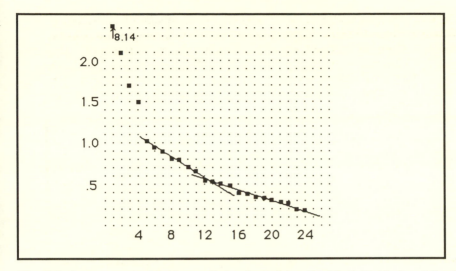

Fig. 5.3 Scree test for Holzinger-Swineford data of Table 5-6. Horizontal axis: eigenvalue number; vertical axis: eigenvalue size.

error factors. The scree test would suggest four factors in this example, for the four eigenvalues rising above the scree.

Figure 5.4 shows the scree test applied to the eigenvalues from random data. In this case, there are no true factors arising above the rubble of the scree, which begins with the first eigenvalue. Again, the scree has an initial, approximately linear segment, and then further out another section of slightly

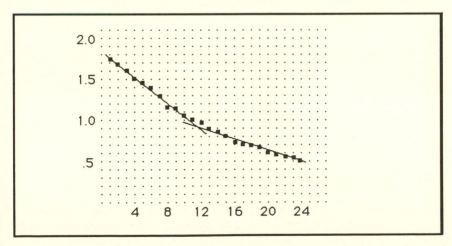

Fig. 5.4 Scree test for random data of Table 5-6. Horizontal axis: eigenvalue number; vertical axis: eigenvalue size.

lesser slope. In this example, the scree test would provide much better guidance to the number of factors than would the Kaiser-Guttman rule--although either approach would work fairly well for the data of Fig. 5.3.

Figure 5.5 applies the scree test to the artificial example of Table 5-1. This illustrates a difficulty of applying the scree test in small problems: There is not enough excess of variables over factors to yield sufficient rubble for a

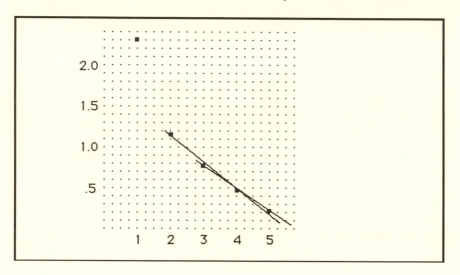

Fig. 5.5 Scree test for data of sample problem of Table 5-1. Horizontal axis: eigenvalue number; vertical axis: eigenvalue size.

well-defined scree. The Kaiser-Guttman rule would suggest 2 factors in this case. A scree test would indicate the presence of at least one real factor and would not be very compelling after that--one could make a case for 1, 2, 3, or more factors. The graph is *consistent* with the presence of 3 factors, but one's confidence in the true linearity of a slope defined with just two points cannot be very high!

Most users of the scree test inspect visual plots of the eigenvalues in the manner we have described. However, a computer-based version also exists (Gorsuch,1983, p.168).

Chi-square test of residuals

As noted earlier in this chapter, the Canonical factors are maximum likelihood factors, that is, each factor represents the best possible fit to the residual correlations, according to a maximum likelihood criterion. This presents the possibility of a χ^2 test after the addition of each factor, as to whether the correlations implied by the factors extracted so far constitute an adequate account of the original correlations. Such a test may be thought of either as a test of the goodness of fit of the model to the data, or as a test for the insignificance of the residuals left when the correlations implied by the model are subtracted from the observed correlations.

Table 5-7 Chi square test of residuals after each factor (sample problem of Table 5-1, with communalities assumed known), N = 100

	Factor pattern		
	I	II	III
D	.06	.35	.18
E	.68	.51	-.12
F	.52	.48	.20
G	.94	-.17	-.01
H	.51	-.22	.24
χ^2	94.67	3.15	0.00
df	10	6	3
p	<.001	>.50	--

Note: Maximum likelihood factors extracted successively from residual matrices via LISREL. Chi square test is for the significance of residuals after the extraction of the given factor (= test of goodness of fit of R_r to $impR_r$ based on all factors so far); df given by Table 5-8 formula + n, since the n communalities are not solved for.

The calculations for the example problem of Table 5-1 are shown in Table 5-7. The maximum likelihood factors were obtained by successive extractions of a single factor from residual matrices using LISREL; as can be seen, they are essentially identical to the rescaled canonical factors calculated via eigenvalues and vectors in Table 5-3. Chi squares were obtained on the assumption that the correlations were based on 100 subjects. As can be seen, a statistical test at a conventional significance level would have concluded that two factors, plus sampling error, provide a plausible explanation of the data-- agreeing, in this case, with the Kaiser-Guttman rule. Only if the expected sampling error were considerably decreased, e.g., if a sample size of upwards of 400 were assumed, would a χ^2 test suggest the extraction of a third factor from this particular matrix of correlations. (Note that such "what if" questions are easily answered in such cases because the χ^2s go up proportionally to N-1.)

Maximum likelihood factors are often obtained in conjunction with an iterative solution for the communalities, as illustrated in Table 5-8. In this approach one solves iteratively, using a maximum likelihood criterion, for

Table 5-8 Maximum likelihood solution with iterated communalities (sample problem of Table 5-1), N = 100

	Factor pattern		
	I	II	h^2
D	.00	.37	.14
E	.56	.56	.63
F	.41	.63	.56
G	1.00	.00	1.00
H	.51	-.06	.26
χ^2	25.90	1.19	
df	5	1	
p	<.001	>.20	

Note: df = 1/2 $[(n-m)^2 - n - m]$, where n = number of variables and m = number of factors (Harman, 1976, p. 205).

the pattern coefficients and communalities for 1,2, . . . , k factors, stopping when a model is reached that is not rejected by the χ^2 test at the prechosen level of significance.

    Such a solution for the example problem of Table 5-1 is given in Table 5-8. The initial communality estimates were squared multiple correlations. A Heywood case was encountered for variable G on the second iteration; its communality was fixed at 1.00 and the iteration allowed to proceed to completion. Again, the χ^2 test after the extraction of the second factor is no longer significant. In this case the next factor would have negative degrees of freedom for a χ^2 test. The dfs for successive factors are fewer in this analysis than in that of the preceding table because the communalities are solved for as well as the pattern coefficients.

Cross-validation

The ultimate test of any method of choosing the number of factors to extract is that it selects factors that will be found again in new samples of subjects and new sets of tests covering the same domain. If the ultimate criterion is cross-validation, why not use it as the immediate criterion? Indeed, several such procedures have been suggested. One such method, which cross-validates across subject samples, has been proposed by Cudeck and Browne (1983).

The method goes as follows:

1. Split the subjects randomly into two equal subsamples, call them A and B.
2. Take one subsample, say A, and factor with increasing numbers of factors, 1,2,3, . . . ,k.
3. After each factor is extracted, apply a goodness-of-fit test, such as maximum likelihood, to the discrepancy between the correlations implied by the factors extracted in A and the observed correlations in the *other* subsample, B.
4. Repeat, factoring in B and testing against A.
5. Take as the optimum number of factors the one that yields the best cross-validation indices of fit. Ideally, this number will turn out to be the same for both directions of cross-validation. If it is not, one could argue either for taking the smaller of the two, or an intermediate number. In Cudeck and Browne's examples (discussed shortly), if the number was not the same in both directions it was usually close.

The fit of the model to the correlations in the original subsample necessarily improves as the number of factors increases. But the fit to the correlations in the opposite subsample typically improves and then deteriorates, suggesting that after awhile the factoring in the first subsample is merely fitting idiosyncrasies of sampling error, making the fit in the second sample get worse instead of better.

Table 5-9 provides some examples from Cudeck and Browne. The data were subsamples drawn from a large study in which six ability tests were given on three occasions to a large group of high school students. There were thus 18 measures intercorrelated for each subsample, from which 9 factors were extracted using a maximum likelihood procedure. After each factor, the maximum likelihood criterion was calculated for the fit to the opposite subsample correlations; these numbers are presented in the table. As you can see, for the smallest subsample size (75 subjects apiece), the best cross-validation in each direction was for a 3-factor solution. For the 200-subject subsamples, 4 or 5 factors represented the best cross-validation, and for 800-subject subsamples, 8 or 9 factors were optimum.

The authors also reported the number of factors that would have been chosen in each case based on a χ^2 test in the original sample. These χ^2s are not shown in the table, but a § symbol marks the smallest number of factors that yielded an acceptable solution (p > .05). For the smaller samples, the number of factors cross-validating tended to be less than than the number that were statistically significant in the original sample (3 factors versus 5 factors for N = 75). For large samples, the two criteria tended to yield comparable numbers of factors. Cudeck and Browne studied other sample sizes as well as those shown; they also demonstrated that their procedure can be used with other structural models besides factor models.

Table 5-9 Using a cross-validation criterion in choosing the number of factors, for three sample sizes (data from Cudeck & Browne, 1983)

Number of factors	N = 75		N = 200		N = 800	
	A	B	A	B	A	B
1	5.02	4.99	2.56	2.52	1.75	1.82
2	4.94	5.14	2.05	2.13	1.31	1.34
3	4.91*	4.71*	1.85	1.87	.92	.97
4	5.30¶	5.05	1.83*	1.90	.81	.81
5	5.55§	5.20§¶	1.90	1.66*	.72	.73
6	5.72	5.37	1.83§¶	1.77§	.69	.68
7	5.97	5.61	1.86	1.74¶	.61	.68
8	5.88	5.79	1.88	1.79	.56*	.64§
9	--	5.70	1.91	1.75	.58¶§	.64*¶

* best cross-validation criterion.
§ smallest number of factors with p > .05.
¶ number of factors chosen by Akaike's criterion.

In addition, they report results for Akaike's information criterion (see Chapter 2) based on the χ^2 in the original sample. The number of factors yielding the most parsimonious solution based on this criterion is shown by the ¶ symbol in Table 5-9. In general, the parsimonious solution by Akaike's criterion corresponded quite well with the first solution acceptable at the .05 level of significance. Both methods slightly overfactored relative to the cross-validation criterion in smaller samples but tended to agree with it in large ones.

Factor Extraction Using Packaged Programs--SPSS, BMDP, SAS, MINITAB

Several widely used statistical program packages contain facilities for doing factor analysis. This section describes the handling of factor extraction by several such programs; their use in factor rotation is considered in Chapter 6. Additional information may be found in a paper by MacCallum (1983), which compares the SPSS, BMDP and SAS factor analysis programs in some detail. Two cautions: Programs are not static entities--they get revised by their makers from time to time; and program packages are not monolithic--sometimes different subprograms within a package do things differently. There is no substitute for a direct test to determine what the version of program X currently on your local computer actually does in situation Y.

Here we consider the SPSS program called FACTOR (Nie, Hull, Jenkins,

Steinbrenner, & Bent, 1975), the SAS program PROC FACTOR (SAS Institute, 1982), and the BMDP program called 4M (Dixon, 1983). These three programs are generally similar in what they do, but not identical. Furthermore, they contain some minor differences in the way they go about doing it that may trip up the unwary user who is familiar with one of them and tries to use another with only a superficial glance at the manual.

In general, SPSS FACTOR might be the simplest to use for a beginner, and SAS PROC FACTOR would probably be the choice of a more sophisticated user who would appreciate the versatility of its many options. BMDP4M tends to lie somewhere between the other two. But in ease of use, the overriding consideration would be familiarity with the general system: Someone who knew other SAS or BMDP or SPSS programs should find the corresponding factor analysis program quite easy to learn and use, whereas someone coming as a complete stranger to a system must acquire a considerable baggage of information about forms of data input, handling of missing values, managing of data and output files, conventions for program representation, etc., common to all the programs in that system. A user might in practice find it easier to accomplish a particular analysis in devious or arcane ways in the system he or she knows, than to learn from scratch another system in which the same analysis can be carried out in a simple and straightforward manner.

In the factor extraction step, all three programs can carry out a simple principal factor analysis of a correlation matrix, with user- supplied communalities. (If these are 1.0, this is a principal components analysis.) BMDP and SAS can also analyze covariance matrices; SPSS cannot. Appendix E shows examples of program setups in the three systems that will accomplish a principal factor analysis for the correlation matrix of Table 5-1, yielding the solution given in Table 5-2.

All three programs can carry out a principal factor analysis with iteration to improve communalities. SPSS starts this process from SMCs (unless the correlation matrix is singular, in which case it automatically switches to the highest r in the row as a starting point). BMDP and SAS allow the user more flexibility. Both allow SMCs or highest r in the row, as well as user-specified values, and SAS has a couple of other options as well (including setting communalities at random!).

All three programs allow control of the number of factors either by the user specifying a fixed number of factors, or declaring a minimum eigenvalue beyond which factoring will not continue. There is, however, a difference in how this is done. SPSS gives priority to the number of factors declaration--i.e., it ignores the minimum eigenvalue specification if NFACTORS is set. SAS and BMDP, however, will use whichever criterion is satisfied first. Thus to insure extracting a given number of factors, a low eigenvalue must be set, or to insure a cutoff at a particular eigenvalue, a large number of factors should be specified. SAS uses "all" and "zero" as the default values for number of factors and minimum eigenvalue, respectively, which usually means that if the user specifies one and not the other, the specified parameter will govern, but BMDP uses 1.0 as the default value for the minimum eigenvalue, so the user who simply designates a given number of factors may well find fewer being extracted.

One further complication in using the minimum eigenvalue cutoff is that SAS and BMDP typically apply this to the eigenvalues of the matrix being factored, which is not the appropriate eigenvalue for the Kaiser-Guttman rule--recall that this refers to eigenvalues of the original correlation matrix **R**, not **R**$_r$. In order to use the Kaiser-Guttman rule in these cases, one must first carry out a principal components analysis to ascertain the appropriate number of factors and then use this number to control the factor analysis. SPSS always bases its minimum eigenvalue cutoff on the eigenvalues of the original correlation matrix and uses 1.0 as the default value for the minimum eigenvalue, making it very easy to apply the Kaiser-Guttman rule.

To do a scree test with BMDP or SPSS, one must obtain the appropriate eigenvalues and plot them by hand or other program; SAS can produce a scree plot as an available option.

SAS offers a considerable array of alternative procedures for factor extraction: alpha factor analysis, maximum likelihood factor analysis, and several other methods, including some we have not discussed, such as image analysis, least squares analysis, and Harris component analysis. SPSS FACTOR offers alpha, canonical, and image analysis, although the canonical program is somewhat suspect (MacCallum, 1983). A separate SPSS program, JFACTOR, does maximum likelihood and least squares. The SPSS-X version of FACTOR combines the two. BMDP offers the least extensive array of additional alternatives for factor extraction: maximum likelihood and image analysis--the latter only as a component of a package called *little jiffy*.

Finally, the three programs differ in what they do if a Heywood case arises during the iteration of communalities. SPSS stops and prints out the results at the point where a communality first reaches 1.0. Depending on the program, BMDP may set the offending communality to 1.0 and continue iterating, or it may quit without printing out any results. SAS allows the user to select from several options for dealing with Heywood cases, including fixing the communality to 1.0, stopping, or permitting the Heywood solution to proceed.

All three of the programs allow the user to begin an analysis from raw data or from a correlation matrix entered by the user or produced by other programs in the series. All permit either proceeding directly to the factor extraction step, or saving the results of the factor extraction for reentry into different rotation procedures.

Packaged factor analysis programs such as these are convenient for use in many practical situations, but there is another approach that allows one complete freedom to experiment with new methods and procedures, or to verify the analyses that existing programs actually carry out (which may sometimes be at variance with what they appear to be from descriptions in the manual, or with what other programs with similar names may do). This approach is to make use of a general-purpose program capable of doing matrix operations.

One can then carry out any sequence of operations that can be stated in the language of matrix algebra, which is the primary language of documentation of factor analysis procedures in the technical literature.

Two languages of this kind might be mentioned. One is MINITAB (Ryan et al., 1985). This is an easy-to-use, undergraduate-oriented language for simple statistical procedures, but it contains matrix facilities as well: It can read, copy and transpose matrices, and execute the elementary matrix operations of addition, subtraction, multiplication, and inversion. It can also obtain the eigenvalues and vectors of a matrix, and so can carry out essentially all of the basic operations involved in the factor extraction phase of factor analysis.

A second option is the MATRIX procedure in SAS. This is a more complex and flexible language, which puts a considerably more extensive set of matrix operations at the disposal of the user (and is consequently more difficult for a beginner to use, particularly one not already at home in SAS).

There are other possibilities as well: For example, the International Mathematical and Statistical Library of computer subroutines (IMSL, 1984) contains many matrix-oriented FORTRAN subroutines that the user can assemble into programs. Clearly, one's horizons need not be limited by the options that the standard factor analysis packages elect to provide.

Several methods have now been presented for arriving at an initial factor solution. In the next chapter we will consider ways in which this initial solution may be transformed so as to simplify its factor pattern.

Chapter 5 Notes

There are a number of excellent books on factor analysis in which you can pursue further the topics of this and the next chapter. Examples include:

Harman, H. H. (1976). *Modern factor analysis* (3rd ed.). Probably the best systematic treatment of the many variants of factor analysis. Extensive worked examples.

Mulaik, S. A. (1972). *The foundations of factor analysis.* A well-regarded general text.

Gorsuch, R. L. (1983). *Factor analysis* (2nd ed.). Readable and up-to-date, with a practical research emphasis. Less formal than Harman or Mulaik.

Rummel, R. J. (1970). *Applied factor analysis.* A text from the perspective of a political scientist (most of the others are by psychologists).

Lawley, D. N., & Maxwell, A. E. (1971). *Factor analysis as a statistical method* (2nd ed.). An emphasis on statistical inference and the use of maximum likelihood methods.

Cattell, R. B. (1978). *The scientific use of factor analysis.* A rich but opinionated treatment by a creative contributor to the field.

Chapter 5 Exercises

Problems 1-3 involve the following correlation matrix **R**:

1.00	.28	-.14	.42
.28	1.00	-.08	.24
-.14	-.08	1.00	-.12
.42	.24	-.12	1.00

1. Obtain eigenvalues and eigenvectors of **R**, using any available computer program that yields normalized eigenvectors (**V´V = I**). Rescale the eigenvalues to principal factor pattern coefficients, **P**, by **VL**, where **L** is a diagonal matrix of the square roots of the eigenvalues. Show that by using 1, 2, 3, and 4 factors (i.e, 1 to 4 columns of **P**) **PP´** gives increasingly accurate reconstructions of **R** (but comment).

2. If **U**^{2} is a diagonal matrix of uniquenesses with elements .51, .84, .96, and .64, obtain **R**$_r$ as **R - U**^{2}. Obtain the eigenvalues and eigenvectors of **R**$_r$, and convert to **P** (set any small imaginary square roots to zero). Use the first column of **P** to reconstruct **R**. Comment.

3. Extract a single centroid factor from **R** using the highest *r* method of estimating the communality (don't forget to reflect). Compare to the first principal factors obtained in Problems 1 and 2.

Problems 4 to 8 involve the following **R** matrix of five socioeconomic variables for a group of Los Angeles census tracts (Harman, 1976, p. 14).

Variable	1.	2.	3.	4.	5.
1. Total population	1.00	.01	.97	.44	.02
2. Median school years		1.00	.15	.69	.86
3. Total employment			1.00	.51	.12
4. Professional services				1.00	.78
5. Median house value					1.00
Standard deviations	3440	1.8	1241	115	6368

4. Estimate the communalities of **R** by squared multiple correlations, using any available matrix inversion program.

5. How many factors would be indicated for **R** by the Kaiser-Guttman rule? By the scree test?

6. Using the communalities estimated in Problem 4, obtain R_r. Convert this to a (reduced) covariance matrix C_r by SR_rS, where S is a diagonal matrix of the standard deviations of the variables. Obtain principal factor pattern matrices P for R_r and C_r via eigenvalues and vectors. Rescale the latter by $S^{-1}P$. Comment.

7. Assuming that the correlations of R were based on 25 census tracts, test the hypothesis using LISREL that a single common factor would fit these data. (Hint: fix TD to U^2 and analyze R matrix.) Would you accept or reject the hypothesis of a single common factor?

8. Using any standard factor analysis computer program package, obtain--if available--(unrotated) principal factors, Alpha factors, and maximum likelihood factors. If your program package permits it (not all do), use SMCs without iteration as communality estimates. (If you use iterative estimation, watch out for Heywood cases.) In any event, compare your results to those obtained in Problem 6.

Chapter Six:
Exploratory Factor Analysis--II.
Transforming the Factors to Simpler Structure

In the last chapter we pursued one approach to simplicity: to account adequately for the data with the smallest number of latent variables, or factors. The strategy was to solve for a series of uncorrelated general factors of decreasing size, each accounting for as much as possible of the covariation left unaccounted for by the preceding factors.

As we noted earlier, the next step is to transform such solutions to simplify them in another way--to minimize the number of paths appearing in the path diagram. This process is what factor analysts have traditionally called *rotation*. It received this name because it is possible to visualize these transformations as rotations of coordinate axes in a multidimensional space. A serious student of factor analysis will certainly want to explore this way of viewing the problem, but we do not need to do so for our purposes here. References to the spatial approach crop up from time to time in our terminology--uncorrelated factors are called *orthogonal* (at right angles), and correlated factors are called *oblique*, because that is the way they looked when the early factor analysts plotted them on their graph paper. But for the most part we view the matter in terms of iterative searches for transformation matrices that will change initial factor solutions into final ones that account just as well for the original correlations but are simpler in other ways. Table 6-1 provides an example, based on a correlation matrix derived from Fig. 6.1 (by now you should be able to verify readily that the path diagram would indeed produce this correlation matrix and the communalities h^2 shown alongside it). The matrix P_0 represents principal factors derived from the eigenvalues and vectors of R_r (using exact communalities), in the manner outlined in the last chapter. It is simple in the first sense we have considered: $P_0 P_0'$ reconstructs R_r exactly (within rounding error); i.e., the two factors account for all the common variance and covariance in the matrix, and the first accounts for as much as possible by itself. The factor pattern is not, however, simple in the second sense. Only one, or at most two,

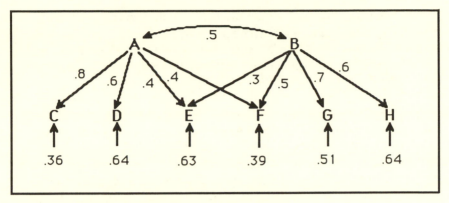

Fig. 6.1 Two factor example to illustrate rotation.

Table 6-1 Example of rotated 2-factor solution (artificial data based on Fig. 6.1; exact communalities)

R

	C	D	E	F	G	H	h^2
C	1.00	.48	.44	.52	.28	.24	.64
D	.48	1.00	.33	.39	.21	.18	.36
E	.44	.33	1.00	.47	.35	.30	.37
F	.52	.39	.47	1.00	.49	.42	.61
G	.28	.21	.35	.49	1.00	.42	.49
H	.24	.18	.30	.42	.42	1.00	.36

P_0

	I	II
C	.704	-.379
D	.528	-.284
E	.607	-.032
F	.778	.073
G	.596	.368
H	.510	.315

T

	A	B
I	.607	.547
II	-.982	1.017

$F = (T'T)^{-1}$

	A	B
A	1.00	.50
B	.50	1.00

P

	A	B
C	.80	-.00
D	.60	-.00
E	.40	.30
F	.40	.50
G	.00	.70
H	.00	.60

Note: **R** = correlation matrix; **P_0** = initial principal factor pattern; **T** = transformation matrix; **P** = transformed factor pattern; **F** = factor intercorrelations.

paths (from the second factor to E and F) are small enough to plausibly be considered negligible.

Next to **P_0** in the table is a matrix **T**. For the moment we will not worry about how it was obtained--by magic, perhaps. But what it does is to produce by the matrix multiplication P_0T a new matrix P, one that has several zero paths, four, in fact, and whose remaining paths--to two decimal places--agree perfectly with those of the model that generated the data. As shown below T, one can also obtain as a function of T the intercorrelation matrix of the factors, F, again in agreement with the model of Fig. 6.1.

The factor pattern P is "just as good as" P_0 in the sense that both can reconstruct the original (reduced) correlation matrix R_r with two factors-- although because the factors represented by P are correlated, we must take this into account. For P_0, we can use P_0P_0' to yield the matrix R_r. With P, we use PFP', where F is the factor intercorrelation matrix. This is the more general formulation and includes P_0P_0' as a special case: because the initial factors are uncorrelated, their intercorrelation matrix is an identity matrix and can be dropped from the expression.

If we know T, then, we can transform P_0 to the simpler pattern P that we seek (assuming that such a simpler pattern exists). How can we find T? In some cases it can be obtained by direct calculation, but in general it is pursued by a process of iterative trial and error, and nowadays a computer usually carries out the search.

A variety of different procedures exist for this purpose, going by such exotic names as Varimax, Quartimax, Oblimin, Orthoblique, and Promax, to name just a few of the more popular ones. (Gorsuch, 1983, gives a table listing 19 such procedures and describes it as a "sample.") We will say something in a moment about the differences among the methods, but for the moment let us consider them as all doing the same thing: modifying some initial arbitrary T (such as an identity matrix) by some form of systematic trial and error so that it yields a P which, while retaining its capacity to reconstruct R_r, gets progressively simpler and simpler in the second sense of containing an increasing number of zero or near-zero paths.

Why are there 19-plus methods extant? Several reasons might be mentioned. First, some procedures incorporate particular constraints. For example, some seek the best solution with uncorrelated factors, others allow factors to become correlated. Some procedures allow a general factor to emerge, others avoid one. Second, the different procedures differ in such practical characteristics as how widely they are available, how expensive they are to run, how robust they are in the face of various adversities, and so on. And finally, none of them works best on all problems. On a particular correlation matrix, method A finds a simple P that method B does not; on another matrix, method B goes to an elegant solution like a hot knife through butter, whereas method A bogs down hopelessly. Of course, on many problems with fairly simple and clear-cut structure, any of a variety of procedures will locate that structure and yield basically similar results, as we shall see in the case of the simple example problem that we have just considered.

Orthogonal Transformations--Quartimax and Varimax

We begin by considering two orthogonal procedures, Quartimax (Neuhaus & Wrigley, 1954) and Varimax (Kaiser, 1958). These two procedures seek for a **T** that will produce factors that are uncorrelated with one another; that is, after the transformation the factors remain independent, but they are simpler in the sense of having more zero or near-zero paths.

Both methods use a criterion of simplicity of **P** that is based on the sum of the fourth powers of the pattern coefficients, and both modify **T** in an iterative fashion until a **P** is reached for which the criterion cannot be improved. In both, the changes in **T** are introduced in such a way that $(\mathbf{T}'\mathbf{T})^{-1}$, the factor intercorrelation matrix **F**, always remains an identity matrix. The criteria used, and hence the properties of the solutions, are, however, a little different.

Quartimax uses as a criterion just the sum of the fourth powers of the elements of **P**: in symbols,

$$\Sigma\Sigma p^4,$$

where p represents a pattern coefficient, and the $\Sigma\Sigma$ means to sum over both rows and columns.

Varimax subtracts from this sum a function of the sum of squared coefficents within columns of **P**. The Varimax criterion may be given as

$$\Sigma\Sigma p^4 - \tfrac{1}{k} \Sigma_f(\Sigma_v p^2)^2,$$

where Σ_f and Σ_v indicate summing across factors and variables, respectively, and k is the number of variables.

The sums of fourth powers of the coefficients in a **P** matrix will tend to be greater when some coefficients are high and some are low than when all are middling (given that in both cases the correlation matrix is equally well reconstructed, and the factors remain orthogonal). Thus, the iterative process will tend to move toward a **P** matrix with a few high values and many near-zero values, if such a matrix can be found that continues to meet the other requirements.

The Quartimax criterion is indifferent to where the high values are located within the **P** matrix--many of them could be on a single general factor, for example. The Varimax modification awards a bonus to solutions in which the variance is spread out more evenly across the factors in **P**, so Varimax tends to avoid solutions containing a general factor.

Either procedure may be, and Varimax usually is, applied to variables that have first been rescaled so their communality equals 1.0; that is, the scaling used in Alpha factor analysis is employed. This tends to prevent the transformation process from being dominated by a few variables of high communality. Varimax applied to variables rescaled in this way is usually referred to as "normal" or "normalized" Varimax--as opposed to "raw" Varimax,

161

in which the criterion is calculated on coefficients in their ordinary scaling. The rescaling is easily accomplished by dividing every coefficient in a row of the factor pattern matrix by the h^2 of that variable before beginning the rotational process, and then scaling back by multiplying by h^2 at the end. This procedure is also sometimes referred to as "Kaiser normalization," after its inventor.

Both Quartimax and Varimax are relatively fast and robust procedures and are widely available in standard computer factor analysis packages. They can be used with confidence whenever conditions are suitable (i.e, where the causal factors underlying the observed correlations are expected to be independent of one another, or nearly so). The choice between them would depend on whether one wishes to encourage or discourage the appearance of a general factor. If in doubt, one can, of course, try it both ways. If they arrive at the same solution, well and good. If they do not, one knows that there is more than one plausible intepretation of the data.

Even when moderately correlated factors are expected, these methods are sometimes still used because of their other virtues. With correlated factors, they cannot be expected to provide very neat solutions, but they will often identify the main factors correctly. If an orthogonal procedure is used when factors are really correlated, the low correlations will only be relatively low, not near zero as they would be with an oblique factor solution, but the high and low correlations will often match fairly well in the two solutions.

Table 6-2 gives examples of Quartimax and Varimax solutions based on the sample problem of Table 6-1. An initial principal factor solution was transformed so as to maximize the Quartimax or Varimax criterion. The raw versions were used to keep the examples simple. Note that the **T**'s for orthogonal rotations are symmetrical, apart from signs.

It will be observed that the Varimax **P** approximates the values of the original path model fairly well in its larger coefficients, but that the small ones are systematically overestimated. The Quartimax **P** assigns relatively more variance to the first factor, making it a fairly general factor.

From the values of the Quartimax and Varimax criteria given at the bottom of the table, you can see that each criterion is highest for its own solution (as it should be). The initial principal factor solution is not too bad by the Quartimax criterion because it does have some high and some low loadings, but it is unsatisfactory to Varimax because the principal factor solution maximizes the difference in variance between the two factors.

In this example, Varimax does better than Quartimax at approximating the paths of the original model, and either one does better than the initial principal factor solution. The advantage Varimax has here results from the fact that the model to be approximated has two roughly equal factors--that is, there is no general factor present.

Varimax and Quartimax can be considered special cases of a general class of orthogonal transformations called Orthomax, whose criterion can be written:

$$\Sigma\Sigma p^4 - w\,^1/_k \Sigma_f (\Sigma_v p^2)^2 \ .$$

The weight w determines the particular criterion. If $w = 0$, the second part of the expression vanishes, and we have the Quartimax criterion. With $w = 1$, we have Varimax. Intermediate values of w would yield solutions with intermediate properties. A negative value of w would award a bonus to solutions with unequal factors, instead of a penalty, and so on.

Table 6-2 Factor pattern matrices, factor intercorrelations and transformation matrices for Quartimax and Varimax transformations of an initial principal factor solution (example problem of Fig. 6-1)

P	Initial		Quartimax		Varimax		Paths	
	I	II	A	B	A	B	A	B
C	.70	-.38	.78	-.17	.78	.20	.80	.00
D	.53	-.28	.59	-.13	.58	.15	.60	.00
E	.61	-.03	.59	.13	.47	.39	.40	.30
F	.78	.07	.73	.28	.52	.58	.40	.50
G	.60	.37	.47	.52	.19	.67	.00	.70
H	.51	.32	.41	.44	.16	.58	.00	.60

F	I	II		A	B		A	B		A	B
I	1.00	.00	A	1.00	.00	A	1.00	.00	A	1.00	.50
II	.00	1.00	B	.00	1.00	B	.00	1.00	B	.50	1.00

T		A	B	A	B
I		.962	.273	.736	.677
II		-.273	.962	-.677	.736

Criteria			
Quartimax	1.082	1.092	1.060
Varimax	.107	.204	.389

Note: Communalities for initial solution iterated from SMCs. Raw Quartimax and Varimax transformations. Paths from path diagram.

Oblique Transformations

When operating in a situation where the true underlying factors are substantially correlated, the procedures of the last section cannot achieve ideal solutions. A number of methods have been proposed for locating good solutions when factors are correlated with one another ("oblique"). Because the factor intercorrelations represent additional free variables, there are more possibilities for strange things to happen in oblique than in orthogonal solutions. For example, two tentative factors may converge on the same destination during an iterative search, as evidenced by the correlation between them becoming high and eventually moving toward 1.00--this cannot happen if factors are kept orthogonal. Despite their real theoretical merits, oblique solutions tend to be more expensive to compute, more vulnerable to idiosyncracies in the data, and generally more likely to go extravagantly awry than orthogonal ones. There is no one oblique procedure that works well in all situations, hence the proliferation of methods. We discuss in detail only two of the many existing methods and mention a few others briefly. First, we consider Direct Oblimin, an approach that pursues a general iterative solution based on improving a criterion as in the Quartimax or Varimax solutions, except that the requirement that factors be uncorrelated is dropped. Then we discuss a procedure, Promax, which attempts to retain the general robustness of the orthogonal methods while arriving at an oblique factor solution.

Direct Oblimin

The criterion used in the Direct Oblimin procedure (Jennrich & Sampson, 1966) is as follows--the criterion is minimized rather than maximized:

$$\Sigma_{ij} \, (\Sigma_v p_i^2 p_j^2 - w \, \tfrac{1}{k} \, \Sigma_v p_i^2 \, \Sigma_v p_j^2).$$

Σ_{ij} refers to the sum over all factor pairs ij ($i < j$), and the other symbols are as used for the Varimax criterion. The weight w (sometimes given as δ) specifies different variants of Oblimin that differ in the degree to which correlation among the factors is encouraged. If $w = 0$, only the first part of the expression, the products of the squared pattern coefficients on different factors, is operative. This variant is sometimes given a special name, Direct Quartimin. It tends to result in solutions with fairly substantial correlations among the factors. By making the weight w negative, high correlations among factors are penalized. Most often, zero weights or modest negative weights (e.g., $w = -.5$) will work best. Large negative weights (e.g., $w = -10$) will yield essentially orthogonal factors. Positive weights (e.g., $w = .5$) tend to produce overoblique and often problematic solutions.

The term *direct* in the title of Direct Oblimin indicates that the criterion is applied directly to the factor pattern matrix **P**. (There also exists an Indirect

Oblimin in which the criterion is applied to another matrix). Again, as in the case of the orthogonal procedures, the scaling of Alpha factor analysis may be used, so that the transformation is carried out on a factor matrix rescaled so that all the communalities are equal to 1.0, with a return to the original metric at the end.

Table 6-3 presents a Direct Oblimin solution for the sample 2-factor problem. (The Quartimin version--w=0--was used.) Observe that the pattern coefficients approximate the paths of the initial model quite well, except that the near-zero loadings are slightly negative. The correlation between factors is a little on the high side (.57 vs. .50), but on the whole the solution has recovered quite well the characteristics of the path diagram that generated the correlations.

Below the **F** in the table is the **T**, the transformation matrix that produces the Oblimin **P** from the **P**$_0$ of the initial solution. It is this **T**, of course, that has resulted from iterative modifications by the computer program until the resulting **P** has as low a score as possible on the Oblimin criterion. In the bottom row of

Table 6-3. Factor pattern matrices, factor intercorrelations and transformation matrices for two oblique transformations of an initial principal factor solution (example problem of Fig. 6-1)

P	Initial I	II	Oblimin A	B	Promax A	B	Paths A	B
C	.70	-.38	.81	-.02	.86	-.10	.80	.00
D	.53	-.28	.61	-.01	.65	-.07	.60	.00
E	.61	-.03	.38	.30	.39	.27	.40	.30
F	.78	.07	.37	.51	.37	.49	.40	.50
G	.60	.37	-.05	.73	-.08	.75	.00	.70
H	.51	.32	-.04	.62	-.07	.64	.00	.60

F	I	II		A	B		A	B		A	B
I	1.00	.00	A	1.00	.57	A	1.00	.65	A	1.00	.50
II	.00	1.00	B	.57	1.00	B	.65	1.00	B	.50	1.00

T			A	B		A	B
I			.575	.555		.587	.513
II			-1.071	1.081		-1.178	1.212

Criterion:
Oblimin   .172   .051   .059

Note: Same initial solution as Table 6-2. Oblimin is Direct Quartimin (w=0), not normalized. Both **T** matrices are for rotation from initial solution. Promax based on raw Varimax to 4th power.

the table are given values of the Oblimin criterion for the initial solution, for the Oblimin solution, and for another solution, Promax, discussed shortly. The Promax solution is reasonably good by the Oblimin criterion, although of course not quite as good as Oblimin; both are much better than the initial unrotated solution.

Promax

The Promax solution (Hendrickson & White, 1964) proceeds in two steps. First a Varimax solution is obtained. Then it is transformed to an oblique solution that has the same high and low loadings, but with the low loadings reduced (if possible) to near-zero values. The second step is done by direct calculation, not iteration, so that if an orthogonal solution can correctly identify the factors, Promax provides an efficient route to an oblique solution.

The second step of a Promax solution is a variant of a procedure called Procrustes (Hurley & Cattell, 1962), which forces a factor pattern matrix to a best least squares fit to a predesignated target matrix. It gets its name from the legendary Greek who forced travelers to fit his guest bed by stretching or lopping them as necessary.

In Promax, the target matrix is obtained by raising the elements of the Varimax-rotated pattern matrix to a higher power--usually the third or fourth--and restoring minus signs if the power is even. By raising the pattern coefficients to a higher power, the low values go essentially to zero, and the high values, although they are lowered, remain appreciable, so the contrast between high and low values is sharpened. For example, at the fourth power all loadings of .26 or less become zero to two decimal places, whereas loadings of .70 and .80 remain .24 and .41.

Call the target matrix P_t. Then an initial transformation matrix T_i is obtained by a least squares matrix solution of an overdetermined set of simultaneous equations:

$$T_i = (P_0'P_0)^{-1}P_0'P_t \, ,$$

where P_0 is the initial unrotated factor pattern matrix. This is the first part of the Procrustes solution. The second part is to rescale T_i to its final form T by postmultiplying it by a diagonal matrix D, chosen to make the factor intercorrelation matrix $(T'T)^{-1}$ have diagonal elements equal to 1.0. The necessary D may be obtained as the square roots of the diagonal elements of $(T_i'T_i)^{-1}$.

Table 6-4 illustrates these steps for the Promax solution of Table 6-3, based on the Varimax solution of Table 6-2. At the left of the table is P_t, obtained from the Varimax solution by raising its elements to the fourth power (because they were all initially positive, no restoration of minus signs is

Table 6-4 Calculation of a Promax solution via Procrustes, from the Varimax solution of Table 6-2

P_t		T_i			D			$P = P_0T$	
A	B		A	B		A	B	A	B
.37	.00	I	.171	.115	A	3.253	.000	.84	-.06
.11	.00	II	-.362	.247	B	.000	4.779	.62	-.04
.05	.02							.37	.30
.07	.11		$(T_i'T_i)^{-1}$			$T = T_iD$		.35	.51
.00	.20		A	B		A	B	-.10	.77
.00	.11	A	10.583	9.947	I	.556	.549	-.09	.66
		B	9.947	22.842	II	-1.176	1.180		

Note: P_t is target factor pattern = Varimax solution of Table 6-2 with elements raised to 4th power. $T_i = (P_0'P_0)^{-1}P_0'P_t$ = transformation matrix before rescaling. D = square roots of diagonal elements of $(T_i'T_i)^{-1}$. T = rescaled transformation matrix. P_0 = unrotated factor pattern. P = rotated factor pattern.

required). In the center of the table, T_i, $(T_i'T_i)^{-1}$, D, and T are successively calculated. At the right, T is applied to the unrotated solution to yield a Promax solution. (The resulting solution is similar but not quite identical to the Promax solution in Table 6-3, which was obtained via a computer package program that calculates its target matrix in a slightly more complicated fashion.)

Other oblique rotation methods

As noted earlier, a large number of different methods for oblique factor rotation have been proposed. Some represent slight variations on those mentioned here; the reader can consult other sources (such as Gorsuch, 1983) for more details and citations to original articles. Two methods operating on rather different principles may be worth mentioning briefly. One is the method called Orthoblique (Harris & Kaiser, 1964). This procedure, like Promax, reaches an oblique solution via an orthogonal rotation, but the strategy is a slightly different one. The first k eigenvectors of the correlation matrix (where k is the desired number of factors) are subjected to an orthogonal rotation (originally, raw Quartimax, although others can also be used). The transformation matrix developed in this step is then rescaled in its rows or columns or both by suitable diagonal matrices, to become the final transformation matrix T, the matrix that transforms an initial principal factor solution into the final rotated oblique solution.

The other principle to be considered is embodied in procedures such as Maxplane (Cattell & Muerle, 1960) and the KD transformation (Kaiser &

Madow,1974). These methods focus specifically on low pattern coefficients and work at transforming factors to get as many pattern coefficients close to zero as possible. Methods such as these strive directly for the second kind of simplicity in a pattern matrix--a large number of near-zero paths. They are most often used to apply final touches to an approximate solution arrived at by another procedure. As a case in point, the KD procedure applied following an Orthoblique rotation yielded a solution in the Table 6-3 rotation problem that was an almost exact reproduction of the underlying path model, differing from it by no more than .01 in any correlation or path coefficient.

Factor pattern and factor structure in oblique solutions

As mentioned previously, two matrices relating the observed variables to the latent variables are frequently reported. One is the factor pattern matrix **P** that we have already discussed. The other is the factor structure matrix **S**, which is a matrix giving the correlations between the factors and the variables. When factors are uncorrelated (orthogonal), there is just a single path from any factor to any variable, and hence the correlation between them is numerically equal to the path coefficient. In this case, therefore, **S** equals **P** and only one need be reported. However, in an oblique solution, there will be additional compound paths between factors and variables via correlations with other factors, and **S** will in general not be equal to **P**. However, **S** may readily be calculated from **P** and the factor intercorrelations **F** by the equation:

$$S = PF.$$

Table 6-5 gives an example for the oblique solution of Table 6-1 (two-factor example with exact communalities). Note that the **S** matrix does not have the zero paths of the **P** matrix; because the factors are correlated, each is

Table 6-5 Factor structure, for example of Table 6-1

	Factor pattern P			Factor intercorrelations F			Factor structure S	
	A	B		A	B		A	B
C	.80	.00	A	1.00	.50	C	.80	.40
D	.60	.00	B	.50	1.00	D	.60	.30
E	.40	.30				E	.55	.50
F	.40	.50				F	.65	.70
G	.00	.70				G	.35	.70
H	.00	.60				H	.30	.60

correlated with variables on which it has no direct causal influence. (The reader should verify from Fig. 6.1 that the matrix multiplication that yields **S** is equivalent to determining the correlations from tracing the paths in the path diagram.)

The reference vector system

When factor analyses were done by hand on desk calculators, the minimization of computations took precedence over conceptual simplicity, and a clever scheme of Thurstone's (1947) that permitted carrying out the factor analysis calculations indirectly on a system of *reference vectors* was widely employed, because it required fewer matrix inversions during the analysis. Although this scheme is not much used nowadays (and if it is, it is likely to be concealed within the computer program), the student needs to be aware of its existence in reading the earlier literature, where results are often presented in this form.

When factors are orthogonal, the distinction between the two systems may be ignored because the factor pattern and structure not only coincide with each other, but also with the pattern and structure expressed in terms of the reference vectors.

When factors are correlated the two systems become distinct, but there is a simple relationship of proportionality between the reference vector structure-- the correlations of variables with the reference vectors--and the factor pattern **P**. Note that this is a relation of structure (correlations) in one system to pattern (path coefficients) in the other. Thurstone's famous criterion of "simple structure," which was the goal of his solutions in the reference vector system, corresponds to simplicity of the pattern matrix **P**, which is the goal of the factor transformation procedures we have described in this chapter.

For the details of calculating the diagonal matrices that will transform results back and forth between the two systems, the reader may consult Harman (1976) or other standard factor analysis texts.

Estimating Factor Scores

Given a factor solution, and the scores of individuals on the observed variables, can we assign to the individuals scores on the latent variables, the factors? This would be attractive to do, if we assume the latent variables to represent fundamental causal influences underlying the interrelations of the superficial measures we actually observe.

The answer is, in general, No we cannot, although we can provide *estimates* of such scores. These estimates may be quite good if the observed variables are strongly related to the latent variables, or quite poor if they are not.

A number of different methods have been proposed and are discussed in standard factor analysis texts such as Gorsuch (1983) or Harman (1976). One simple one is to add together with equal weight the scores on the observed variables that are most highly correlated with the factor--a robust approach that has a good deal to be said for it. However, the most widely used method is to recognize that we are dealing with a prediction situation, in which we want to predict the latent variable, the factor, from the set of observed variables. An accepted way of making predictions of a given variable from a set of related variables is by carrying out a multiple regression.

Recall that in multiple regression one solves for a set of weights (called "beta weights"), which can be applied to the observed variables to predict the unknown variable. To solve for such weights, one needs to know the correlations among the predictor variables, and the correlations of these with the variable being predicted. Then the vector of beta weights may be obtained by premultiplying the latter by the inverse of the former. The matrix of correlations among the predictors (the observed variables) is of course obtainable--it is just the correlation matrix **R**. The correlations of the observed variables with a factor is a column of the factor structure matrix **S**, let's call this **s**. So we can get the desired beta weights **b** for estimating a factor as follows:

$$\mathbf{b} = \mathbf{R}^{-1}\mathbf{s} \ .$$

These weights **b**, applied to the observed variables in standard-score form, will yield the best prediction, in the least squares sense, of this factor from these variables. The equation

$$\mathbf{B} = \mathbf{R}^{-1}\mathbf{S}$$

will give the beta weights, as columns of **B**, for the whole set of factors simultaneously.

One can work with raw instead of standard scores by rescaling the betas appropriately and supplying an additive constant, but we need not deal with such complications here. Any standard text on multiple regression will supply the details.

All the other paraphernalia of regression analysis apply. The vector multiplication $\mathbf{b}'\mathbf{s}$ gives the square of the multiple correlation of the factor with the predictors and thus represents the proportion of variance of the factor that is predictable from the observed variables. This will give one some notion of how well the factor scores are estimated in the present set of data. To get an idea of how well these factor estimation weights will transfer to new samples, a cross-validity coefficient can be calculated (Rozeboom, 1978):

$$R_c^2 = 1 - (1 - R^2)(N + m)/(N - m) \ .$$

The R^2 inside the parentheses is the squared multiple correlation, N is the sample size, and m is the number of predictors (observed variables). The

corrected value R_c^2 is an estimate of how well the beta weights calculated in the given sample will predict in the population (and hence, on the average, in new random samples from that population). If the measurement is good--that is, the obtained multiple correlation is high--and if the ratio of subjects to variables is large, one would not expect much falling off of prediction in a new sample. For instance, in a 6-variable problem based on 100 subjects in which the obtained multiple correlation is .90, the expected drop off when using the factor estimation weights in a new sample is only to a correlation of .89:
$R_c^2 = 1 - (1 - .81) \ 106/94 = .786; \sqrt{.786} = .89.$ If one were to do a 17-variable factor analysis on 50 subjects and obtain a multiple correlation of .70, the expected drop would be all the way to zero (try it in the formula and see), and the factor score estimation would be completely worthless.

The factor scores estimated by regression using beta weights and standardized predictors will have a mean of zero and a variance equal to R^2. If it is desired to produce the factor scores in standard score form, which is customary, the beta weights can simply be multiplied by 1/R before applying them to the (standardized) predictors. (A moment's reflection should show why this is so: If it doesn't, ask yourself, What is the standard deviation of the initial set of estimated factor scores? and reread this paragraph.)

Table 6-6 (next page) illustrates the calculation of factor scores for our two-factor example. At the top of the table are R^{-1}, the inverse of the correlation matrix, and **S**, the factor structure matrix (the latter from Table 6-5). $R^{-1}S$ yields the beta weights, **B**. **B′S** yields a matrix that has the squared multiple correlations from the regressions in its principal diagonal; the reciprocals of their square roots constitute the rescaling matrix $D^{-1/2}$, which rescales **B** to **W**. **W** contains the coefficients that produce standardized estimates of the factor scores.

The final step is taken in the bottom row of the table. Hypothetical data for three subjects on the six variables are shown. Postmultiplied by **W**, these yield the estimates (in standard-score form) of scores for these three individuals on the two factors A and B.

The correlation among these factor estimates can be obtained, if desired, by pre and postmultiplying **B′S** by $D^{-1/2}$. The estimates for the two factors turn out to be correlated .67, somewhat higher than the correlation of .50 between the factors themselves. This is a typical result when estimating factor scores by this method.

Table 6-6 Estimating factor scores by regression (two-factor example of Tables 6-1 and 6-5)

R^{-1}	Inverse of correlation matrix							S	Factor structure	
	C	D	E	F	G	H			A	B
C	1.64	-.48	-.32	.52	.01	.00		C	.80	.40
D	-.48	1.36	-.14	-.22	.00	.00		D	.60	.30
E	-.32	-.14	1.43	-.31	-.18	-.12		E	.55	.50
F	-.52	-.22	-.31	1.88	-.48	-.33		F	.65	.70
G	.01	.00	-.18	-.48	1.44	-.35		G	.35	.70
H	.00	.00	-.12	-.33	-.35	1.32		H	.30	.60

B	Beta weights				W	Factor score weights	
	A	B	**B'S**			A	B
C	.51	-.01	.769	.502	C	.58	-.01
D	.22	-.00	.502	.725	D	.25	-.00
E	.15	.13			E	.17	.15
F	.23	.35	**$D^{-1/2}$**		F	.27	.41
G	-.00	.38	1.140	.000	G	-.00	.44
H	-.00	.26	.000	1.174	H	-.00	.30

Z	Data (standard scores)						Z_F	Factor scores	
	C	D	E	F	G	H		A	B
	1.2	.6	1.5	.8	.1	1.1		1.31	.92
	-1.0	-1.6	-.1	.0	.8	-1.4		-.99	-.07
	-.7	1.2	.9	-1.0	-1.3	.7		-.23	-.64
	.	.	.	.	.	.		.	.
	.	.	.	.	.	.		.	.

Note: $B = R^{-1}S$; $D = \text{diag } B'S$; $W = BD^{-1/2}$; $Z_F = ZW$.

Factor score indeterminacy

It is tempting to interpret the factor score estimation problem as though there were a "real" set of factor scores out there somewhere, and our difficulty is in not being able to estimate them accurately. But in a sense, the fundamental problem is not really one of estimation, it is that a given factor solution (**P**, **S**, and **F**) just doesn't define factors uniquely. For any given **P**, **S**, and **F** there is a range of potential factors that are equally compatible with the obtained results. If the communalities of the original variables are high, these potential factors will tend to be highly correlated, much like one another, and the choice among them may not be crucial. But if the communalities are low, some potential factors may actually have zero correlations with others (McDonald & Mulaik,

1979). In short, if we hope to score subjects accurately on latent variables, they should be latent variables with strong and preferably multiple links to data. If there are important aspects of our latent constructs that are not well reflected in our measurements, and many aspects of our measures unrelated to the latent constructs, we should not be surprised if there is ambiguity in trying to assess the one from the other.

Extension analysis

Suppose we carry out an exploratory factor analysis on a set of variables, and have available the scores of additional variables for the same subjects. Can we extend the factor solution to these new variables? The answer is, Yes we can, but subject to the same limitation as with calculating scores on the factors.

The simplest way, conceptually, is just to obtain the estimated scores on a factor and correlate these scores with the scores on the additional variables. This will provide the matrix S for the new variables, their correlations with the (estimated) factors. One can then obtain the factor pattern matrix P for the new variables by the relationship $P = SF^{-1}$, the inverse of the relationship $S = PF$ given earlier for getting the factor structure from the factor pattern matrix. Again, this is the factor pattern for the new variables with respect to the estimated factors, and the utility of an extension analysis thus is dependent to a considerable degree on the presence of conditions that minimize factor indeterminacy and lead to accurate estimation of factor scores.

In practice we do not actually have to go through the step of calculating factor scores for individuals--a matrix shortcut exists. To obtain the estimated correlations of a factor with the new variables, one may multiply the matrix of correlations of the new variables with the old ones, call it Q, times the vector of beta weights scaled to produce standardized factor scores, call it w; that is,

$$s = Qw.$$

If W is a matrix with rescaled beta weights as its columns, the equation becomes:

$$S = QW,$$

providing a convenient way of calculating S, and thence P.

Table 6-7 illustrates the extension of the example two-factor analysis to two new variables, I and J. Hypothetical correlations of the two new variables with the original six variables are shown as Q, as well as the factor score coefficients W from the preceding table. The factor structure matrix S for the new variables is obtained by the matrix multiplication QW, and the factor pattern matrix P by SF^{-1}, where F^{-1} is the inverse of the factor intercorrelation matrix F from Table 6-1. The matrix S gives the estimated correlations of the new variables with the two factors A and B, and P gives estimates of the paths

Table 6-7 Extension of factor analysis of Table 6-1 to two new variables I and J

Q	Correlations of new variables with original variables						W	Factor score weights	
	C	D	E	F	G	H		A	B
I	.60	.40	.50	.60	.50	.40	C	.58	-.01
J	.20	.10	-.10	-.30	-.10	-.20	D	.25	-.00
							E	.17	.15
S	Factor structure			P	Factor pattern		F	.27	.41
	A	B			A	B	G	-.00	.44
I	.69	.66		I	.48	.42	H	-.00	.30
J	.05	-.24		J	.22	-.36			

Note: **S = QW**; **P = SF^{-1}**. Factor score weights from Table 6-6.

from the factors to the new variables.

There is a question that might have occurred to some readers: If one has scores on the additional variables for the same subjects, why weren't these variables just entered into the factor analysis in the first place, yielding correlations and path coefficients with the factors directly? There could be several reasons why one might not do this. The additional scores might only have become available after the factor analysis was carried out. Or the additional variables might have been excluded from the factor analysis to avoid distorting it or biasing it in some way; for example, some variables might have been excluded because of artifactually high correlations with included variables, or because they were external reference variables which were desired for help in interpreting the factors, but which one did not want influencing the factor analysis itself. Or one might have an extremely large number of variables available, only a subset of which could feasibly be used for a factor analysis, but all of whose relationships with the factors would be of interest. In any of these situations, an extension analysis could be the answer.

Factor Rotation Programs--SPSS, BMDP, and SAS

In the last chapter, the factor extraction procedures contained in a number of statistical program packages were considered. In this section we briefly examine the procedures for rotation that these packages offer.

All three packages provide the basic orthogonal rotation procedures Varimax and Quartimax; BMDP and SAS also provide the more general

Orthomax procedure of which these two are special cases.

There is more variation among oblique rotation procedures. SPSS offers only one, Direct Oblimin, which BMDP also provides, though SAS does not. SAS, on the other hand, offers both Promax and Procrustes, which the others don't. SAS has Orthoblique as well; BMDP has it only as part of the separate "little jiffy" package, and SPSS does not have it at all.

SAS and BMDP permit either Kaiser's normalization of variables prior to factor rotation, or rotation of the raw factors; SPSS automatically normalizes.

The packages differ in their default philosophy--i.e., in what they do if no specification is made concerning rotation. SPSS and BMDP rotate automatically, using their default rotation procedure, which in both cases is Varimax. In order not to rotate, the user must specify a no-rotation option. In SAS the default is for no rotation, so if the user wishes to rotate he or she must specify a particular rotation procedure.

All the rotation methods described in this chapter, except Procrustes, are iterative in character and so would be quite tedious to carry out by simple matrix calculations, as in MINITAB. However, one can go a long way with one basic iterative orthogonal routine, say Varimax, and a matrix package. Such oblique procedures as Promax and Orthoblique do their basic iterative rotation orthogonally and then transform to an oblique solution at the end.

An Example: Thurstone's Box Problem

In this section, we carry through an exploratory factor analysis from raw scores to factor scores. We use as an example a demonstration problem originally devised by Thurstone (1947) and later modified by Kaiser and Horst (1975). The intent is to illustrate exploratory factor analysis in a situation in which the true underlying latent variables are known, so we can check our results.

Thurstone began with a hypothetical population of 20 rectangular boxes; the first three are illustrated at the top of Table 6-8. For each box, a number of "scores" were derived, as mathematical functions of one or more of its three dimensions length (= X), width (= Y), and height (= Z). Thus, the first box, which was 3 units long by 2 wide by 1 high, had a score on the first variable (X^2) of 9, on the second (Y^2) of 4, and so on. The 4th, 5th, and 6th variables are products of two dimensions each, the 7th through 9th are natural logarithms, and the 10th is the triple product XYZ. (Thurstone actually created 20 variables, but we are using only 10 of them to keep the example more manageable.)

Kaiser and Horst added 5% random error to Thurstone's scores, in the interest of greater realism, and doubled the number of boxes from 20 to 40 by using each twice (with different random errors). The resulting scores for the first 3 boxes, rounded to 1 decimal place, are shown in the columns labeled *observed* in Table 6-8. The complete data matrix is given in Appendix F. The matrix of score intercorrelations is given in Table 6-9, and the eigenvalues and a scree test in Table 6-10 and Fig. 6.2. Clearly, by either the Kaiser-Guttman

Table 6-8 Scores on 10 variables for first 3 boxes, Thurstone's box problem (data from Kaiser and Horst, 1975)

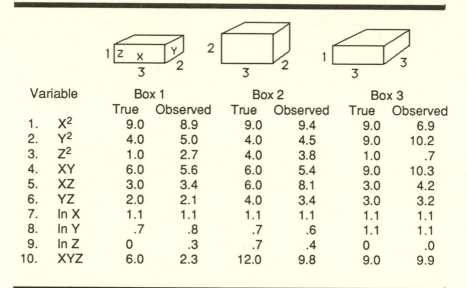

Variable	Box 1 True	Box 1 Observed	Box 2 True	Box 2 Observed	Box 3 True	Box 3 Observed
1. X^2	9.0	8.9	9.0	9.4	9.0	6.9
2. Y^2	4.0	5.0	4.0	4.5	9.0	10.2
3. Z^2	1.0	2.7	4.0	3.8	1.0	.7
4. XY	6.0	5.6	6.0	5.4	9.0	10.3
5. XZ	3.0	3.4	6.0	8.1	3.0	4.2
6. YZ	2.0	2.1	4.0	3.4	3.0	3.2
7. ln X	1.1	1.1	1.1	1.1	1.1	1.1
8. ln Y	.7	.8	.7	.6	1.1	1.1
9. ln Z	0	.3	.7	.4	0	.0
10. XYZ	6.0	2.3	12.0	9.8	9.0	9.9

Table 6-9 Correlation matrix, Thurstone box problem

	V1	V2	V3	V4	V5	V6	V7	V8	V9	V10
V1	1.00	.23	.08	.64	.42	.15	.92	.10	.10	.46
V2		1.00	.17	.84	.21	.57	.37	.93	.13	.53
V3			1.00	.14	.85	.84	.16	.27	.91	.77
V4				1.00	.34	.50	.73	.76	.12	.65
V5					1.00	.78	.50	.22	.87	.87
V6						1.00	.30	.64	.81	.91
V7							1.00	.28	.18	.59
V8								1.00	.21	.58
V9									1.00	.76
V10										1.00

rule or the scree test, a 3-factor solution is indicated. Table 6-11 shows three principal factors based on the (unrounded) correlation matrix, using iterated communality estimates starting from SMCs. As you can see, the communality estimates are all quite high, in the range .90 to .97, consistent with the fact that the variables contain around 5% random error.

176

Table 6-10 Eigenvalues, Thurstone box problem

1.	5.621	6.	.079
2.	2.348	7.	.072
3.	1.602	8.	.049
4.	.100	9.	.025
5.	.088	10.	.015

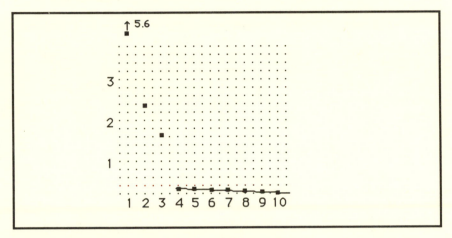

Fig. 6.2 Scree test for Thurstone box problem.

Table 6-11 Unrotated principal factors, Thurstone box problem

Variables	F1	F2	F3	h²
V1	.503	.468	.683	.938
V2	.649	.506	-.499	.925
V3	.732	-.615	-.020	.914
V4	.736	.647	-.056	.963
V5	.834	-.400	.282	.934
V6	.904	-.280	-.252	.959
V7	.633	.470	.558	.933
V8	.664	.382	-.604	.951
V9	.717	-.627	.045	.909
V10	.971	-.108	.047	.956

Table 6-12 shows two transformed solutions and a direct confirmatory maximum likelihood solution. The first solution is an orthogonal rotation using Varimax. For most practical purposes, this would be quite adequate: it correctly identifies the three factors as respectively reflecting the latent dimensions Z, Y, and X that underlie the manifest measurements, as one may verify by comparing the rotated factor pattern to the path diagram of Fig. 6.3.

Table 6-12 Final factor solutions, Thurstone box problem

	Varimax			Direct Oblimin			Confirmatory		
	F1	F2	F3	F1	F2	F3	F1	F2	F3
V1	.08	.06	.96	-.00	-.08	.99	.00	.00	.96
V2	.11	.94	.16	-.03	.96	.03	.00	.96	.00
V3	.95	.08	-.02	.98	-.04	-.11	.95	.00	.00
V4	.10	.77	.60	-.05	.72	.51	.00	.70	.53
V5	.90	.05	.36	.90	-.12	.29	.85	.00	.36
V6	.84	.50	.05	.80	.41	-.09	.79	.48	.00
V7	.17	.22	.93	.07	.08	.92	.00	.00	.95
V8	.20	.95	.02	.08	.98	-.12	.00	.95	.00
V9	.95	.02	.02	.99	-.10	-.07	.95	.00	.00
V10	.79	.43	.39	.73	.30	.28	.71	.36	.36
F1	1.00	.00	.00	1.00	.28	.22	1.00	.15	.11
F2		1.00	.00		1.00	.29		1.00	.25
F3			1.00			1.00			1.00

Note: Kaiser normalization used for Varimax and Oblimin solutions.

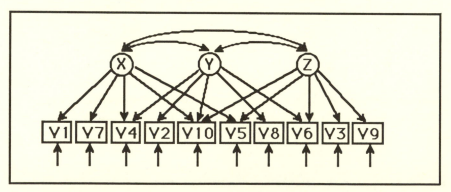

Fig. 6.3 Path diagram of Thurstone box problem.

Nevertheless, because nearly all the minor loadings in the Varimax solution are positive, there is an indication that the underlying dimensions X, Y, and Z are slightly correlated with one another; i.e., that there is a general size factor in the population of boxes. Therefore, an oblique solution (Direct Oblimin; *w* = 0) was carried out; it is also shown in Table 6-12. This yields a slightly cleaner solution; whereas the Varimax factor pattern had several of its minor loadings in the .15 to .25 range, the oblique solution has all its near-zero loadings .12 or less in absolute value. The correlations among the factors are modest--in the .20s. There is, however, a suggestion that this solution may be a little too oblique: 11 of the 15 near-zero loadings are negative.

On the far right in Table 6-12 is a confirmatory maximum likelihood analysis of the correlation matrix via LISREL, based on the path diagram of Fig. 6.3: the zero values shown were fixed, and the nonzero values solved for. Note that this solution is a trifle less oblique than the Oblimin solution but agrees with it in finding the three dimensions to be slightly correlated in these data, suggesting the presence of a (modest) general size factor. We can compare the factor intercorrelations with the true correlations among the length, width, and height dimensions, calculated for Thurstone's original population of boxes: $r_{XY} = .25$, $r_{YZ} = .25$, $r_{XZ} = .10$. Obviously, the two oblique factor solutions have come reasonably close--the Oblimin, as suspected, has slightly overestimated the correlations.

Finally, Table 6-13 shows factor scores for the first three boxes, from the Oblimin solution. Comparison to standardized values of the true scores suggests that, with these high communalities, the factor scores do a reasonable (although not a perfect) job of estimating the true-score values.

Table 6-13 Theoretical true standard scores compared to factor scores, first three boxes (Oblimin solution)

	Original			Standardized			Factor scores		
	X	Y	Z	X	Y	Z	F3	F2	F1
Box 1	3	2	1	-1.43	-1.29	-1.17	-1.47	-1.05	-.93
Box 2	3	2	2	-1.43	-1.29	.13	-1.26	-1.40	-.26
Box 3	3	3	1	-1.43	.00	-1.17	-1.42	.13	-1.05

Note: Means for original scores over population of 40 boxes: 4.1, 3.0, 1.9; SD's: .768, .775, .768; factors reordered for ease of comparison.

Thus, a factor analysis *can* correctly recover information about known latent variables, even in the face of a certain amount of random error and nonlinearities of relationship between the latent and manifest variables. However, a caution: It is not always this easy. This is, after all, an example in which three and only three major latent dimensions are present. In most real-life data sets confronting social and behavioral scientists, there are likely to be numerous lesser causal factors in addition to the few major factors the analyst is interested in. The smaller of these extraneous factors can usually be safely lumped together under the heading of random error--this means, incidentally, that the communalities are likely to be rather lower than the .95's of the box example. But in addition there are likely to be some appreciable nuisance factors present, not quite large enough to isolate successfully, yet sufficiently large to distort systematically the picture presented by the main variables. The investigator may be deemed fortunate who encounters a situation as clear-cut as Thurstone's boxes.

Chapter 6 Notes

Many references to the research literature using exploratory factor analysis can be found in the texts listed at the end of Chapter 5. Harman (1976, pp. 7-8), for example, cites studies in fields as diverse as economics, medicine, geology, meteorology, political science, sociology, biological taxonomy, anthropology, architecture, human engineering, communication, and the study of lightplane accidents. If you want to get really serious, Hinman and Bolton (1979) give short descriptions of approximately 1,000 factor analytic studies published in the areas of experimental, animal, developmental, social, personality, clinical, educational, and industrial psychology during the period 1971-1975!

Other model data sets. Other sets of data with known underlying properties that can be used to try out exploratory factor analysis methods include populations of real measured boxes (Thurstone, 1947, p. 369; Gorsuch, 1983, p. 11), chicken eggs (Coan, 1959), and cups of coffee (Cattell & Sullivan, 1962). An elaborate artificial data set is provided by Cattell and Jaspers (1967).

Rotation criteria. Some computer packages calculate these criteria in forms that are slightly different from but equivalent to those given here. For example, in orthogonal rotations BMDP minimizes a squared cross-products criterion rather than maximizing the sum of fourth powers, but these lead to the same end result.

Factor scores. For a discussion of some alternative approaches, see Saris, de Pijper, and Mulder (1978) and ten Berge and Knol (1985).

Chapter 6 Exercises

Given the initial pattern matrix P_0 below, for two tests of verbal ability and two tests of mechanical aptitude, and the transformation matrix T:

P_0		I	II		T		A	B
	V1	.8	.2			I	.75	.36
	V2	.8	.3			II	.96	-1.16
	M1	.6	-.4					
	M2	.4	-.4					

1. Calculate the rotated factor pattern P. Obtain the factor intercorrelations F, and the factor structure S.

2. Draw the path diagrams for the rotated and unrotated factors, omitting any paths less than .10 in absolute value.

3. Calculate the communalities from the path diagram for the rotated factors, and as the sum of the squares of the rows of P_0. Comment.

4. Using any standard factor analysis program, carry out a Varimax and either a Direct Oblimin ($w=0$) or Orthoblique rotation of P_0. Use Kaiser normalization. Compare the orthogonal and oblique solutions to each other and to the Problem 1 solution.

5. Verify by direct matrix calculation (PFP') that both rotated solutions imply the same R_r, and that this is the same as that implied by the unrotated matrix P_0.

6. Using the Varimax solution of Problem 4 raised to the 4th power as a target matrix, carry out the Procrustes matrix calculations. Obtain P and F for this Promax solution and compare to the oblique solution of Problem 4.

7. Using a computer package program, rotate the following initial factor pattern matrix to oblique solutions, with (normalized) Direct Oblimin, and w values of -1.0, 0, and .5. Comment.

P			
.511	-.104	.252
.426	-.082	.252
.853	-.311	-.031
.587	.479	-.041
.666	-.300	-.275
.542	.501	-.043

8. Estimate factor score weights for the rotated Quartimin ($w = 0$) factors of Problem 7, using a computer program and the regression method. (This may require first reconstructing **R** via **PP′**.) What would the factor scores be of a person who was one standard deviation above the mean on variables 1, 2, and 3, at the mean on 4 and 5, and one standard deviation below the mean on 6?

9. Extend the analysis of Table 6-7 to incorporate a third new variable, whose correlations with variables C through H are .00, -.20, .00, .80, .50 and .00, respectively.

10. Repeat the analysis of Thurstone's box problem, using a different method of factor extraction and a different oblique rotation procedure. Compare your results with those given in the chapter.

Chapter Seven:
Extensions of Latent Variable Analysis

In this chapter we consider several examples of how latent variable analysis can be extended in various ways to handle additional kinds of problems. Some of these developments are still quite recent, and how useful they will prove to be in the long run is therefore not yet clear. Others, such as multidimensional scaling, are well established. All of them, however, represent directions in which latent variable analysis can move beyond the topics and methods we have so far discussed in this book. The sections of this chapter are largely independent. Many readers may elect to browse among them according to their interests. However, most should read the first, and none should omit the last.

Models Involving Means

In the discussion so far, models have been fitted to correlations and to variances and covariances. But they may be fitted to means as well (e.g., Sörbom, 1974).

This is most often done when a model is being fitted in several subpopulations, as was discussed in Chapter 4. If the mean of a latent variable in a particular subpopulation is, say, x units above the grand mean for the total population, the regression coefficients in the model predict by what fraction of x units the means of the observed variables in this subpopulation will be above or below average. Looked at from the other end: Subpopulation differences among the means of observed variables may allow us to make inferences about latent variable means and path coefficients.

Figure 7.1 provides a simple example. Assume we have observed the correlations shown on the right, and that they are the same in the two subpopulations, designated Groups 1 and 2. Suppose that we have also observed the means shown below the path diagrams (given in standard scores based on the combined groups). By the method of triads we can infer from the correlations the path coefficients $a = .6$, $b = .4$ and $c = .8$, in both groups.

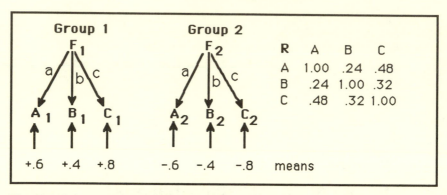

Fig. 7.1 Inferring latent variable means.

Then from the path coefficients and the observed means, because $aF_1 = A_1$, etc., we can infer the mean of F_1 to be 1.0 and the mean of F_2 to be -1.0. (There is an arbitrariness of sign here--we could alternatively assume *a, b,* and *c* to be negative and the signs of F_1 and F_2 to be reversed. This amounts merely to arbitrarily reversing the direction of scoring the latent variable.)

In this example, the means are perfectly consistent with the correlations (mean F_1 is the same whether inferred from mean A_1, B_1, or C_1), and the correlations are identical in the two groups. But, of course, this need not always be the case. As usual, what we can do then is to carry out a least squares or maximum likelihood model fitting to the whole data set, means and covariances in both groups, to get simultaneous best estimates of both the path coefficients and the latent variable means. Any particular assumed values of the unknown latent variable means and path coefficients imply values for the observed means and covariances, and the fitting program iteratively adjusts the former until the discrepancies between the implied and the observed values are minimized.

The fitting process can be carried out by writing equations for the observed values in terms of the unknown parameters, and solving them by a general-purpose fitting program, such as IPSOL. The solution can also be carried out via LISREL by extending the model to include an extra latent variable with zero variance (see Jöreskog & Sörbom,1984, for details).

Tables 7-1 and 7-2 provide an example, based on the path model of Fig. 7.2. The variables differ in variance in the two groups; they are scaled to unit variance in the first group.

Table 7-2 gives a solution based on the assumption of equal path coefficients and error variances in the two groups (one could, of course, make other assumptions). The table shows the results yielded by an iterative solution of the path equations by a general-purpose program using a least squares criterion, and by LISREL with a maximum likelihood criterion. Aside from reversals of sign, the two solutions are essentially similar. The sign reversals

184

indicate that the programs have found two different but equivalent solutions, in which the difference on B between the groups is reversed, as are the signs of the paths from B to the other variables. In effect, the two solutions are assuming latent variable B to be scored in opposite directions, and the various

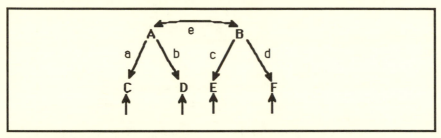

Fig. 7.2 Path diagram of a model to be fit in each of two groups.

Table 7-1 Data for path solution involving means in two groups

		Observed covariances and means								
	Group 1 (N = 100)					Group 2 (N = 120)				
	C	D	E	F	Mean	C	D	E	F	Mean
C	1.00	.50	.20	-.20	5.0	6.00	4.50	1.00	-1.50	14.0
D		1.00	.10	-.20	5.5		4.00	.50	-1.00	12.0
E			1.00	-.20	7.5			1.50	-1.00	4.5
F				1.00	5.5				2.00	10.0

Table 7-2 Solutions for means and variances in two groups, and common paths (based on Fig. 7.2 and data of Table 7-1)

	Paths			Means			Variances			Residual variances	
	LS	ML		LS	ML		LS	ML		LS	ML
a	.88	.88	A_1	-5.6	-5.6	A_1	1.00	1.00	C	.14	.18
b	.65	.64	A_2	4.0	4.6	A_2	7.73	7.58	D	.64	.60
c	.40	-.39	B_1	4.6	-4.2	B_1	1.00	1.00	E	.86	.83
d	-.61	.58	B_2	-3.4	3.5	B_2	3.88	4.13	F	.59	.62
e_1	.44	-.44									
e_2	.48	-.51									

Note: LS = least squares solution, fit by general-purpose program; ML = maximum likelihood solution, by LISREL. Subscripts 1 and 2 designate groups.

185

relationships are simply reversed to reflect this.

The same principles can obviously be extended to situations involving more than two groups. Examples may be found in Sörbom (1974) and Jöreskog and Sörbom (1984).

Multivariate Path Models

All path models are multivariate, of course, in the sense that several variables are involved, but in the present section the term refers to path models in which instead of a single variable at each point, there are several.

Consider Fig. 7.3: On the left is a simple path model in which for some trait mothers and fathers are correlated, and both influence their children (shown here as a son and a daughter, for concreteness). This will, of course, lead to a correlation between sons and daughters for that trait. By the usual path rules (taking the general, unstandardized case), the covariance between sons and daughters is:

$$C_{SD} = aV_Mb + cV_Fd + aC_{MF}d + cC_{FM}b,$$

where the Cs and Vs refer to covariances and variances, and the other symbols are as identified in the figure.

Suppose, however, that each individual was measured on two traits, rather than one. Now the path diagram, shown on the right in Fig. 7.3, becomes considerably more complicated. The four between-generation paths *a* through *d* on the left become the thicket of 16 paths *a* through *p* on the right, and there are 6 mother-father covariances to contend with at the top of the diagram instead of just one. With more than two traits, of course, things get considerably worse. With three traits there would be 36 straight and 15 curved arrows, and with four traits, 64 and 28.

Writing the corresponding path equations gets to be a bit of a nightmare.

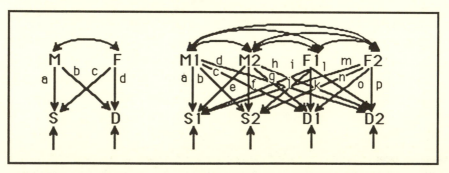

Fig. 7.3 Univariate and bivariate path diagrams. (M, F, S, D = Mother, Father, Son, Daughter; 1, 2 = two traits.)

As a sample, consider just one of the four son-daughter covariances in the two-variable case, C_{S1D1} (the others are C_{S1D2}, C_{S2D1} and C_{S2D2}). From the path diagram:

$$C_{S1D1} = aV_{M1}c + eV_{M2}g + iV_{F1}k + mV_{F2}o$$
$$+ aC_{M1M2}g + eC_{M2M1}c + aC_{M1F1}k + iC_{F1M1}c + aC_{M1F2}o$$
$$+ mC_{F2M1}c + eC_{M2F1}k + iC_{F1M2}g + eC_{M2F2}o + mC_{F2M2}g$$
$$+ iC_{F1F2}o + mC_{F2F1}k .$$

The other three equations are comparable. As one might suspect, things get much worse as the number of variables increases--for the three-variable case there are 9 equations, each more than twice the length of the one above. The four-variable case has 16 equations, with 64 terms apiece. You may not even want to think about the five-variable case--upwards of 2,000 compound paths are involved in the equations.

As Vogler (1985a,b) has pointed out, one need not go to all this trouble-- one can get by with just the simple path diagram on the left of Fig. 7.3, if one replaces the single variables by matrices. Instead of our original equation:

$$C_{SD} = aV_Mb + cV_Fd + aC_{MF}d + cC_{FM}b ,$$

we have:

$$C_{SD} = A'C_MB + C'C_FD + A'C_{MF}D + C'C_{MF}'B .$$

The paths a, b, etc. have simply been replaced by the matrices **A**, **B**, etc., and the single variances and covariances by covariance matrices. Going backwards along an arrow is represented by the transpose of the matrix for going forward (note that this applies to curved arrows as well).

The reader may wish to carry out algebraically the first matrix multiplication on the right-hand side of the preceding expression and convince him or herself that it is yielding as its upper left-hand element terms from the lengthy expression for C_{S1D1}. Use:

$$\mathbf{A} = \begin{matrix} a & b \\ e & f \end{matrix} \quad \mathbf{B} = \begin{matrix} c & d \\ g & h \end{matrix} \quad \text{and} \quad \mathbf{C_M} = \begin{matrix} V_{M1} & C_{M1M2} \\ C_{M2M1} & V_{M2} \end{matrix}$$

(i.e., the single path a from mother to son on the left in Fig. 7.2 becomes the matrix **A** of four paths a, b, e, and f between mother and son in the two-trait diagram on the right.)

It is important to realize that expressing the paths as matrices does not really decrease the number of calculations or the number of parameters involved, it merely simplifies the writing of the equations that generate the expected values from the trial values of the parameters. This is, of course, not a trivial simplification: It is the difference between doing it and not doing it, if one is dealing with more than a couple of variables in a moderately complex path model. And the chances of error in writing the path equations are also vastly decreased if much of the complexity is tucked away into the inner workings of standard matrix multiplication routines.

Further simplifications may be possible. In some cases it may be plausible to assume **A**, **B**, **C**, and **D** to be diagonal: That is, a given parental trait is assumed to influence only the corresponding trait in the child. This considerably reduces the number of free parameters (unknown paths that must be solved for) and thus improves the chances of arriving at robust solutions with reasonable amounts of data. One can still further simplify if, for example, one assumes that fathers and mothers are alike in their influences on their children, or that such influences are the same for both sons and daughters. Naturally, the plausibility of such simplifications will depend on the particular variables involved. In general, it should be possible to test the acceptability of the assumptions by comparing goodness-of-fit χ^2s with and without them.

Rice, Fulker, and DeFries (1986) have applied this method to data from adopted and control families from the Colorado Adoption Project. Their path diagrams for the single-variable case involved 3 observed variables (measures on mother, father, and child) in the control families, and 4 or 5 in the adoptive families (child, two adoptive parents, and either one or both birth parents). There were twice those numbers of latent variables (a genotype and environment for each person). With four manifest variables for each individual-- composite measures of verbal ability, spatial ability, memory, and perceptual speed--it would have been a truly daunting undertaking to write out the individual path equations for solution by their general-purpose model-fitting program. But the substitution of matrices for individual variables made it quite practicable to carry out the analysis.

It is perhaps worth noting that we have encountered examples of this general multivariate principle earlier in this book. Take a simple case from factor analysis. Consider the path diagram of Fig. 7.4, in which X is a

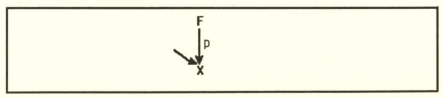

Fig. 7.4 A single-variable, single-factor factor analysis model.

variable, F is a factor, and p is a factor pattern coefficient (a path coefficient). The communality of X, i.e., the variance caused by F, may be obtained as

$\cdot p \; s_F{}^2 \; p$. But suppose we had several X's and several F's? We just substitute matrices. The expression for the implied variance-covariance matrix of the X's becomes the familiar **PFP′**. Given standardized latent and observed variables, **F** is the factor intercorrelation matrix, **P** is the factor pattern matrix, and the whole expression yields the implied intercorrelation matrix $\mathbf{R_r}$.

In this, and other such cases, the generalization from the single-variable to the multivariate situation is direct.

Nonlinear Effects of Latent Variables

The relationships expressed in path models are linear. Path models are, after all, a special application of linear regression. However, it is well known that in linear regression one can express nonlinear and interactive relationships by the device of introducing squares, products, etc. of the original variables. Thus, to deal with a curvilinear prediction of Y from X we might use the prediction equation:

$$Y = aX + bX^2 + Z .$$

Or we could deal with an interactive effect of X and Z on Y with an equation such as:

$$Y = aX + bZ + cXZ + W .$$

These equations represent nonlinear relationships among observed variables by the use of linear regressions involving higher order or product terms. Can the same thing be done with *latent* variables? Kenny and Judd (1984) explore this question and conclude that the answer is Yes. We follow their general strategy.

Suppose we wish to represent the first relationship in the preceding paragraph, but X is an unobserved, latent variable, indexed by two observed variables, call them A and B. In structural equation form:

$$A = aX + U$$
$$B = bX + V$$
$$Y = cX + dX^2 + Z .$$

The first two equations constitute the measurement model, the third the structural model. (For simplicity we are treating Y as an observed variable, but it could be a latent variable as well, with its own indexing measures.)

But how is X^2 to be linked to the data? Kenny and Judd point out that the preceding equations imply relationships of X^2 to A^2, B^2 and the product AB. For example, by squaring the equations for A and B we obtain the first two of the following equations, and by taking the product of the equations for A and B we obtain the third:

$$A^2 = a^2X^2 + 2aXU + U^2$$
$$B^2 = b^2X^2 + 2bXV + V^2$$
$$AB = abX^2 + aXV + bXU + UV .$$

Figure 7.5 represents these various relationships in the form of a path diagram. Notice that X, X^2, XU, and XV are shown as uncorrelated. This will be

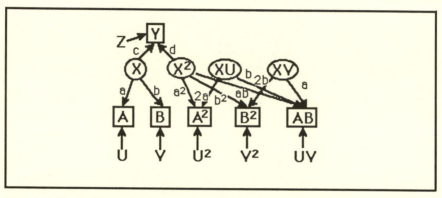

Fig. 7.5 Path diagram for nonlinear effect of X on Z.

Table 7-3 Path equations based on path diagram of Fig. 7.5

Path equations		
If:		
$V_X = x$	$V_A = x + u$	$V_{A^2} = 2x^2 + 4xu + 2u^2$
$V_U = u$	$C_{A,B} = bx$	$C_{A^2,B^2} = 2b^2x^2$
$V_V = v$	$C_{A,A^2} = 0$	$C_{A^2,AB} = 2bx^2 + 2bxu$
$V_Z = z$	$C_{A,B^2} = 0$	$C_{A^2,Y} = 2x^2d$
Then:	$C_{A,AB} = 0$	$V_{B^2} = 2b^4x^2 + 4b^2xv + 2v^2$
$V_{X^2} = 2x^2$	$C_{A,Y} = xc$	$C_{B^2,AB} = 2b^3x^2 + 2bxv$
$V_{U^2} = 2u^2$	$V_B = b^2x + v$	$C_{B^2,Y} = 2b^2x^2d$
$V_{V^2} = 2v^2$	$C_{B,A^2} = 0$	$V_{AB} = 2b^2x^2 + b^2xu + xv + uv$
$V_{XU} = xu$	$C_{B,B^2} = 0$	$C_{AB,Y} = 2bx^2d$
$V_{XV} = xv$	$C_{B,AB} = 0$	$V_Y = c^2x + 2d^2x^2 + z$
$V_{UV} = uv$	$C_{B,Y} = bxc$	

Note: C = covariance, V = variance; path *a* is set to 1.0.

190

the case if X, U, and V are normally distributed and expressed in deviation score form. Kenny and Judd also show that given these assumptions, expressions can be derived for the variances of the square and product terms. Under these conditions the following relations hold:

$$V_{X^2} = 2V^2_X \; ; \; V_{XU} = V_X V_U \, ,$$

and similarly for V_{U^2}, V_{UV}, etc.

This means that we can write equations for the observed variances and covariances of A, B, A^2, B^2, AB, and Y in terms of a moderate number of parameters. If we set to 1.0 one of the paths from X to an observed variable, say *a*, we have left as unknowns the paths *b*, *c*, *d*, the variance of X, and the variances of the residuals U, V, and Z. The remaining values can be obtained from these. The path equations are given in Table 7-3.

Kenny and Judd present illustrative variances and covariances for simulated data from a sample of 500 subjects. These are given in Table 7-4. We have 6x7/2 = 21 observed variances and covariances to fit using 7 parameters, so there are 14 degrees of freedom for the solution.

Table 7-4 Covariance matrix of observed values (Kenny & Judd simulated data, N = 500)

	A	B	A^2	B^2	AB	Y
A	1.150					
B	.617	.981				
A^2	-.068	-.025	2.708			
B^2	.075	.159	.729	1.717		
AB	.063	.065	1.459	1.142	1.484	
Y	.256	.166	-1.017	-.340	-.610	.763

We cannot, however, do the fitting with LISREL. We wish to fix paths and variances in such relations as *b* and b^2, or *b* and 2*b*, and LISREL does not provide for other than EQUAL relationships. Incidentally, we would not want to use a maximum likelihood criterion either, for some of our variables are definitely not normally distributed. We have assumed, for example, that X is normal, but that means that X^2 will certainly not be.

Kenny and Judd fit their model with the program COSAN (McDonald, 1978) mentioned in Chapter 2 that allows the user to specify different relationships among paths. They used a generalized least squares fitting criterion that does not assume multivariate normality.

Their solution is shown in Table 7-5, along with the values used to generate the simulated data, and a solution of the path equations using IPSOL

191

Table 7-5 Solutions of path equations of Table 7-3 for data of Table 7-4

Parameter	Original	IPSOL	COSAN
b	.60	.61	.62
c	.25	.26	.25
d	-.50	-.49	-.50
x	1.00	1.01	.99
u	.15	.15	.16
v	.55	.56	.54
z	.20	.19	.20

Note: COSAN solution from Kenny and Judd (1984)

and ordinary least squares. (The program was the same as that given in Appendix B, except that a higher level of precision was used--the .001s in subroutine MINI were changed to .0001-- and MAXFN was increased to 5000.) Both the COSAN and IPSOL solutions essentially recover the values used to generate the data.

Kenny and Judd go on to carry out a similar analysis for the case of an interactive relationship between two variables, as represented in the second equation given at the beginning of this section. The general principles involved are the same. A path diagram for an example is shown in Fig. 7.6; as you can see, each of the latent variables X and Z is indexed by two observed variables, and there are a number of additional product and residual terms. The 9 observed variables provide 45 variances and covariances, and there are 13 parameters to be solved for (the variances of X and Z, their covariance *i*, the paths *g, h, c, d,* and *e*, and the residual variances S, T, U, V, and W).

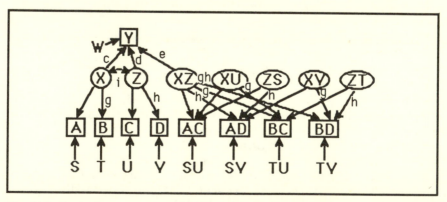

Fig. 7.6 Path diagram for interactive effect of X and Z on Y. Unlabeled paths set to 1.0.

192

Again, Kenny and Judd were reasonably successful in recovering the values used to generate their simulated data.

Obviously, the possibility of constructing and solving path models of nonlinear and interactive relationships broadens considerably the range of latent variable problems that can be dealt with. It may be expected, however, that models involving nonlinear relationships will tend to be fairly demanding in the quantity and quality of data that are required in order to arrive at dependable solutions.

Multidimensional Scaling

Multidimensional scaling (MDS) represents an approach to the discovery of latent variables that is in many ways analogous to exploratory factor analysis. MDS is much younger than factor analysis-- its major development can be dated from Torgerson's 1958 book, *Theory and Methods of Scaling*, although there were some earlier pioneering efforts (e.g., Richardson, 1938). By now, MDS has developed into a substantial methodological subdiscipline of its own. This section only provides a brief account, drawing mostly on Kruskal and Wish's (1978) readable introduction. The interested student may find many further details and examples there or in Davison's (1983) text.

Both MDS and factor analysis can be thought of as engaged in an exploratory search for a small number of latent variables that can provide a framework for explaining a much larger number of relationships among empirical observations. Indeed, both methods, applied to data in a given domain, might very easily come up with the same set of latent explanatory variables. But their approaches would look quite different.

Factor analysis typically begins with the correlations (or covariances) among a set of observed variables and tries to locate a smaller set of latent variables or dimensions that can explain these correlations. MDS typically begins with a set of similarities and dissimilarities among a set of *objects* and looks for dimensions that could account for these. Thus, factor analysis is applied in a two-step process: (1) measure some objects on observable characteristics, and (2) find latent variables to explain the interrelationships among these measurements; whereas MDS goes directly from resemblances among objects to latent variables. (Note that this two-step process of factor analysis does not refer to the stages of factor extraction and rotation we have discussed earlier, but to two stages underlying the factor extraction itself.)

Suppose, for example, that one is interested in the underlying dimensions of personality in a certain population. A typical approach of a factor analyst would be to give subjects from that population a number of personality tests and inventories, and then to carry out an exploratory factor analysis of the intercorrelations among the personality scales. A multidimensional scaler would, however, begin with some overall measure of similarity among people in the sample: Friends might be asked, for example, to rate how similar Joe is to Bill, or Bob to Fred. A multidimensional scaling procedure would then be applied to these similarity judgments to infer the basic dimensions of resemblance among the people--which might, of course, turn out to be the same as the factors discovered by the exploratory factor analysis.

A second difference between MDS and factor analysis is that factor analysis typically assumes ordinary measurement of observed variables, and linear relationships among them, whereas MDS often makes weaker assumptions about its data--say that relationships are merely monotonically, rather than linearly, increasing or decreasing, or that similarities are ordered, rather than on an interval or ratio scale. For the moment, we will stick to metric examples of MDS. Later, we look at the nonmetric variety.

Let us begin with a purely hypothetical experiment, illustrated in Table 7-6. The task is to discover the dimensions of resemblance among the tastes of soft drinks. To keep matters simple, we suppose that only four soft drinks are used. Blindfolded subjects are given all possible pairings of the four drinks and are asked to judge for each pair how dissimilar the two drinks are in taste. Their average dissimilarity ratings are given at the top of the table--the values labeled "d." The ratings mean that Coca-Cola and Sprite tended to be judged quite different, for example, and Coca-Cola and Dr Pepper less so.

MDS begins by considering possible one-dimensional models that might underlie the judgments of differences. One such model is given as #1, where C, 7, D, and S stand for the four soft drinks, arranged in that order along a continuum, with a distance of 2.0 between each adjacent pair. The values δ_1 below the model represent the distances between all possible pairs of soft drink implied by this model.

How do the values δ_1 compare to the values d? If the model is a good representation of the structure of resemblance among the soft drinks, the discrepancies between implied and observed values of dissimilarities should be small. A conventional measure of such discrepancy used in MDS is called *stress*. Stress is based on the sum of squared discrepancies between observed and implied values. The smaller the discrepancies, the smaller the stress, and the better the fit of model to data. It may be helpful to think of stress as an indication of the amount of distortion required to make the model fit the data exactly.

Several different formulas may be used in MDS to evaluate stress. A simple one is:

$$\text{stress} = \left[\frac{\Sigma(d - \delta)^2}{\Sigma d^2} \right]^{\frac{1}{2}}$$

where d refers to an observed dissimilarity, δ to the corresponding implied dissimilarity, and the summing is over all pairings of objects. The Σd^2 in the denominator serves as a scaling factor to allow readier comparison of studies with different numbers of objects, and the final square root returns the measure to the original units of dissimilarity (rather than their squares).

Model #1 in Table 7-6 implies a difference of 2.0 between Coca-Cola and 7-Up, the observed dissimilarity is 4.0, so this pair contributes $(4.0-2.0)^2 = 4.0$

Table 7-6 Example of multidimensional scaling of differences in taste among four soft drinks (hypothetical data)

	7-Up	Dr Pepper	Sprite	
		d		
Coca-Cola	4.0	3.0	5.0	
7-Up		5.0	3.0	$\Sigma d^2 = 100$
Dr Pepper			4.0	

	7-Up	δ_1	Sprite	
Coca-Cola	2.0	4.0	6.0	
7-Up		2.0	4.0	$\Sigma(d - \delta)^2 = 20.0$
Dr Pepper			2.0	Stress = .447

	7-Up	δ_2	Sprite	
Coca-Cola	5.0	2.0	7.0	
7-Up		3.0	2.0	$\Sigma(d - \delta)^2 = 12.0$
Dr Pepper			5.0	Stress = .346

	7-Up	δ_3	Sprite	
Coca-Cola	4.0	3.0	5.0	
7-Up		5.0	3.0	$\Sigma(d - \delta)^2 = 0$
Dr Pepper			4.0	Stress = 0

toward $\Sigma(d - \delta)^2$. Continuing, we obtain a total $\Sigma(d - \delta)^2$ of 20, implying a stress value of .447.

Below this is another model, #2, with a somewhat different ordering and spacing that implies a better fit: its $\Sigma(d - \delta)^2$ is 12, yielding a stress value of .346. Better fits than this are still possible with one-dimensional models--you might care to try your hand at finding one--but no solution in one dimension has a stress anywhere close to zero. Fitting such models by hand via trial and error gets tedious, so, as you may suppose, iterative computer programs have been written to do this. Such programs attempt to find an unimprovable fit of a *k*-dimensional model, starting with $k = 1$. If the stress for the best one-dimensional model remains appreciable, models with two dimensions are tried, and so on, until a satisfactory fit is reached.

In our example, the program by the time it reaches the model labeled #3 has given up on one-dimensional models and begun on two-dimensional ones. We see that now it is able to achieve a perfect fit. The two-dimensional model shown perfectly captures the resemblances among all pairs of soft drinks in the study.

One can then, of course, attempt to determine the meaning of the dimensions--for example, the horizontal dimension might represent a latent variable of cola versus citrus flavor, and the vertical dimension sweetness or fruitiness or some other attribute. (A larger number of soft drinks in the study could help here.) The extremity of an object along a spatial dimension represents its "loading" on that dimension, i.e., it corresponds to the size of a pattern coefficient in factor analysis.

Transformations akin to the factor analyst's rotation are also possible in MDS: One could define a diagonal Dr. Pepper-versus-7-Up dimension in the diagram, and a second, opposite dimension on which Sprite and Coca-Cola would have substantial and opposite loadings. One might question in this instance whether any interpretability had been gained by this change, but in actual practice such rotations are sometimes helpful.

One advantage of MDS is that its starting point, the similarity-dissimilarity matrix among objects, may be obtained in a variety of different ways--including being produced by the judgments of a single subject. This is unlike most applications of factor analysis, which require a substantial number of subjects as the basis for calculating correlations (or covariances) among variables. A few of the methods which have been used to produce initial dissimilarity matrices for MDS are:

(1) Direct judgments of similarity (as in the soft drinks example).

(2) Confusions--two objects are dissimilar if they are rarely mistaken for one another, similar if they often are. This technique has been used, for example, with speech sounds and Morse code signals.

(3) Co-occurences in classification. Judges can be asked to sort objects into categories in a variety of different ways. The frequency with which a pair of objects wind up in the same categories can be used as a measure of their similarity.

(4) Correlations or covariances between objects calculated across some set of attributes, or distances between their profiles across a set of measures.

There are a number of different more-or-less widely available computer programs that do various forms of MDS. Kruskal and Wish (1978, pp. 79-80) give a table listing 14 of them.

Individual differences MDS

The fact that a single individual can yield the similarity matrix for applying MDS opens up a powerful new possibility--to analyze a number of such one-person similarity matrices simultaneously. Such a process is exemplified in the program INDSCAL (Carroll & Chang, 1970). This program yields a group consensus judgment, as does ordinary MDS, but also allows for differences among individuals in the weights they assign to the underlying dimensions.

Figure 7.7 provides an example, based on our hypothetical study of flavors of soft drinks. Let us suppose that the models under (a) on the left represent the results from carrying out the model-fitting process described

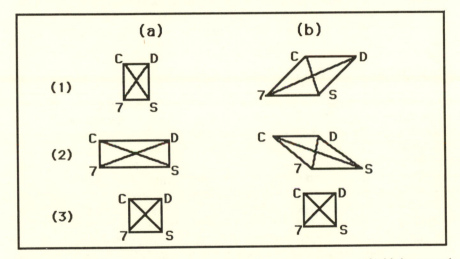

Fig. 7.7 Individuals' models of a common property space--soft drink example of Table 7-6.

earlier, using the similarity judgments of three individual subjects. Each shows the same general kind of two-dimensional structure among the flavors, but the individuals differ in the weight given to the two underlying dimensions in arriving at their overall similarity judgments. The individual labeled (1) tends to emphasize the cola-citrus dimension more than the average judge does, whereas the person labeled (2) puts more emphasis on the second, horizontal dimension. Individual (3) gives nearly equal weight to the two, placing him closer to the group consensus.

The figures under (b) on the right of Fig. 7.5 represent another possible outcome of MDS with individual differences. Suppose that the true underlying dimensions of taste judgment were not the vertical and horizontal ones of the Table 7-6 figure as it stands, but the rotated ones of Dr Pepper versus 7-Up and Coca-Cola versus Sprite. From the group judgments you could not tell this, because the position of the spatial configuration in Table 7-6 is essentially arbitrary. But from individual differences MDS you could, because if the dimensions that the subjects were actually using and to which they were assigning different weights were the diagonal ones in the original figure, the individual subject configurations would tend to look like those on the right-hand side of Fig. 7.5: differently proportioned and slanted parallelopipeds rather than tall or squat rectangles. Individual (1) here is someone who gives more weight to the Dr Pepper-7-Up dimension; individual (2) emphasizes the Coca-Cola-Sprite dimension; person (3) is again more like the group consensus, with roughly equal weight to each.

Thus, individual-differences MDS provides a different basis for a choice out of alternative sets of possible latent variables in an exploratory analysis. In factor analysis, the choice is based on a large number of zero pattern coefficients. In individual-differences MDS, the choice is based on the dimensions along which individual variation in judgments tends predominantly to occur.

Nonmetric MDS

So far, we have supposed that the dissimilarities, both observed and implied, were expressed on quantitative, or metric, scales. This is by no means a necessity of the method--in fact, many users of MDS prefer a nonmetric approach. Such a procedure can be based on weaker and perhaps more realistic assumptions about the underlying processes. It may, for example, only require that the observed and implied dissimilarities among objects fall in the same rank order, rather than being in exact quantitative relation to one another.

For example, judges might not be asked to rate the similarities of soft drink tastes on a numerical scale but rather to make judgments only of greater or less: Is the difference between the taste of Coca-Cola and Sprite greater or less than the difference between the taste of Sprite and 7-Up? From a series of such judgments one can arrive at a rank order of the dissimilarities among the objects being judged; the MDS program then attempts to produce a model in $1, 2, \ldots, k$ dimensions that will come as close as possible to matching that rank order, with closeness again being defined by a suitable measure of stress (indeed, one can use the same formula based on the sum of squared discrepancies as in the metric case--except that the discrepancies are now discrepancies between observed and implied ranks rather than between observed and implied distances).

Table 7-7 provides an example of a problem using ranks. Let us suppose the following experiment. A girl is asked to make judgments about various

Table 7-7 An example of MDS using ranks

	Betsy	Chris	Dana	
		d		
Ann	2	4	6	
Betsy		1	3	$\Sigma d^2 = 91$
Chris			5	

#1

A   B   C   D
6. 5. 7.

	Betsy	Chris	Dana	
		δ_1		
Ann	2	4	6	
Betsy		1	5	$\Sigma(d - \delta)^2 = 8$
Chris			3	Stress = .296

#2

	Betsy	Chris	Dana	
		δ_2		
Ann	2	4	6	
Betsy		1	3	$\Sigma(d - \delta)^2 = 0$
Chris			5	Stress = 0

Note: Distances (5, 6, 7, etc.) are arbitrary; only their ranks count.

pairings of four of her friends: Is the difference between Betsy and Chris greater than that between Ann and Dana? Is Ann more different from Betsy than Chris is? And so on, until it is possible to arrive at the set of ranks given in the table, which is arranged so that 1 means the most similarity and 6 the greatest difference. In the example, our subject sees Betsy and Chris as most alike, and Ann and Dana as most different.

The MDS computer program now seeks a way in which the girls can be arranged along a single dimension so that the implied rank ordering of the distances among them corresponds as closely as possible with the observed rank ordering. The actual distances are arbitrary; it is only their relative size that determines the rankings, and therefore that matters. The solution marked #1 in the table represents one possible arrangement. The rankings derived from it fit pretty well, but not perfectly, to the observed rankings, with a stress value of .296. Although slightly better one-dimensional solutions may be found in this case (you might enjoy looking for one), there is none with a stress of zero. However, in two dimensions (see model #2) arrangements are readily found that yield a perfect match of implied and observed rank orderings.

This provides evidence that our subject judges the resemblances among her friends along at least two latent dimensions, and we can get clues to the nature of these dimensions by such means as having her describe the girls who occupy extreme positions on them; what is there about Dana, for example, that sets her off from Ann and Chris? In a realistic study we would probably want to have our subject rank a somewhat larger number of her friends in order to obtain a more stable definition of the dimensions she uses, as well as to allow for the possibility of identifying more than two dimensions.

The fact that this example has involved a single subject is arbitrary. We might have had each of the girls rank differences among all the rest and then combined their rankings for a single group consensus. The dimensions of judgment would then be those common to the group members. Or we might take an individual-differences approach and consider the rankings by each girl separately. The various possibilities and interpretations would parallel those of metric MDS.

Nonlinear Factor Analysis

Ordinary factor analysis, like the other path models that have been discussed in the earlier chapters of this book, assumes linear relationships between the latent variables--the factors--and the observed, measured variables. What if the relationships in some real-world case are not linear?

If they are nonlinear but monotonic--i.e., the two change together in a constant direction, though not necessarily in equivalent amounts--the variables will agree in the sense of rank order, and a nonlinear version of multidimensional scaling, as described in the last section, could be used. Or, in fact, an ordinary linear approach will often give a decent first approximation to a nonlinear but monotonic relationship.

But suppose the relationship is nonmonotonic, say, an inverted-U function of the kind that may hold between motivation and complex performance, where increasing levels of motivation up to a point improve performance and therefter detract from it. What then?

This issue has been addressed by R. P. McDonald (1962,1967), who notes that in an ordinary factor analysis of the correlations among variables that are related to a latent variable by a curvilinear function of this type, one will tend to obtain two factors. But how can you distinguish the two factors obtained in this case from the two factors obtained when there are two latent variables and ordinary linear relationships?

McDonald's suggestion: Obtain scores for individual subjects on the two factors (actually, for technical reasons he prefers to use principal component rather than factor scores). Plot these scores on a scatter diagram. If the two sets of obtained component scores really reflect a single underlying variable curvilinearly related to the observed measurements, the plotted points should tend to fall along a curved line representing the relationship.

Let us consider an example. Suppose that we have several observed variables Y that are related to a latent variable X by equations of the general form:

$$Y = aX + bX^2 + c.$$

(Recognize the similarity to the case discussed earlier in this chapter of nonlinear relationships among latent variables--there, however, the nonlinearities were in the structural model and the measurement model was assumed to be linear; here we are considering a nonlinear measurement model.)

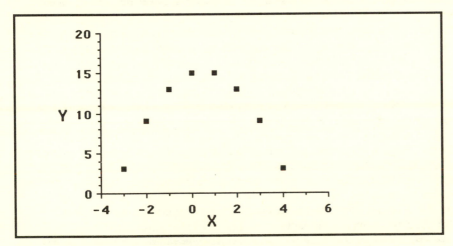

Fig. 7.8 Graph of curvilinear relationship ($a = 1$, $b = -1$, $c = 15$).

The preceding equation specifies a curvilinear relationship between X and Y. For example, suppose that $a = 1$, $b = -1$, $c = 15$, and X varies from -3 to +4 in integer steps. Fig. 7.8 shows the resulting curve. The linear correlation between X and Y is zero, but there is a perfect nonlinear relationship.

Table 7-8 shows a hypothetical example of five tests, Y1 to Y5, each of which is related to the underlying variable X by a quadratic equation of the type mentioned; the tests differ, however, in the relative strength and sign of the linear and quadratic components, and in the contribution of the unique component c.

Table 7-8 Equations to produce scores on tests Y1 to Y5 from given values of latent variable X and specific variables C1 to C5

$$Y1 = X + 2X^2 - C1$$

$$Y2 = 4X - X^2 + 2C2$$

$$Y3 = -X - 3X^2 + C3$$

$$Y4 = -2X + .5X^2 - 2C4$$

$$Y5 = -3X + X^2 + C5$$

For illustrative purposes, 100 simulated subjects were assigned scores on these five tests, by drawing for each subject six random integers in the range ±5, one representing X and one each of the 5 C's, and inserting them in the formulas for the five Y's. These scores were then intercorrelated and two factors extracted by a standard factor analysis program (SPSS Factor), using 1's in the diagonal to yield principal components. The correlation matrix, the pattern coefficients, and the eigenvalues are given in Table 7-9.

Notice that by the Kaiser-Guttman rule this is a very clear two-factor structure. Figure 7.9 shows a scatterplot of the scores on component 1 plotted against component 2. The curvilinear trend of the points is evident. The orientation of the parabolic curve on the graph is arbitrary, since it depends on

Table 7-9 Principal components analysis of the intercorrelations of five hypothetical tests on 100 subjects

---

	\multicolumn{5}{c}{Correlations}					\multicolumn{3}{c}{Factor Pattern}		
	Y1	Y2	Y3	Y4	Y5	C1	C2	h^2
Y1	1.00	-.20	-.98	.17	.47	.75	-.65	.99
Y2		1.00	.23	-.55	-.82	-.74	-.53	.83
Y3			1.00	-.19	-.50	-.77	.63	.99
Y4				1.00	.61	.64	.51	.68
Y5					1.00	.90	.28	.89

Eigenvalues: 2.92, 1.45, .49, .13, .02

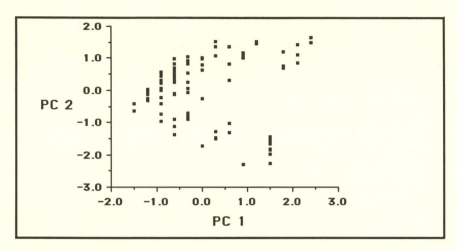

Fig. 7.9 Scatterplot of first two principal component scores from factor analysis of Table 7-9.

just how the two factors emerge, and this will vary with the particular constitution of the tests. McDonald discusses methods of rotating the configuration to a standard orientation, and fitting a parabola to the data, but we need not pursue these matters here.

McDonald also discusses more complex possible cases. For example, a three-factor solution might reflect three ordinary linear latent variables, or one linear and one quadratic relationship, or two linear variables and their product, or first, second, and third powers of a single variable. Although such cases can in principle be handled by the present approach, in practice the discrimination among these alternatives would often place considerable demands on the quality and quantity of available data. Fortunately for the simplicity of life, the variables that social and behavioral scientists measure are most often linearly or at least monotonically related to the underlying latent variables, so that linear methods will normally serve at least as a first approximation. But not always--so if you are working in a domain in which you suspect that nonmonotonic relationships might be present, it would probably not be a bad idea to calculate some principal component scores and do a little plotting. One caution: This method will work better with variables of fairly high communality. With variables of low communality, the amount of scatter of scores due to specificity and error is likely to make it difficult to distinguish any systematic trends in the data. If a latent variable is really only very weakly related to observed variables, establishing the exact form of that relationship may not be easy.

Higher Order Factors

One of the products of an analysis into oblique factors is the matrix **F** of correlations among the factors. This is an intercorrelation matrix, and intercorrelation matrices can be factored. Such a factor analysis of the intercorrelations among factors is called a *second-order* factor analysis. If this factor analysis is also oblique, there will be a matrix of intercorrelations among the second-order factors, which can in turn be factored. This would be called a *third-order* factor analysis. If the third-order factor analysis is oblique . . . . And so on. In principle, this process could go on indefinitely, provided one started with enough variables, but in practice second-order factor analyses are fairly uncommon, third-order factor analyses are decidedly rare, and fourth-order factor analyses are practically unheard of.

Because second- and third-order factor analyses are just factor analyses of correlation matrices, they can be carried out by the same methods used for first-order analyses, with the same issues involved: estimation of communalities, number of factors, orthogonal or oblique rotation, and so on. (A decision at any stage for orthogonal rotation terminates the sequence, of course.) Factor methods involving statistical tests, such as maximum likelihood, should probably be avoided for higher order analyses, because the statistical rationales based on sample size are derived for the case of first-order correlations or covariances and would be of doubtful applicability to the factor intercorrelation matrices involved in higher order analyses. However, causal model-fitting methods, as described in earlier chapters, can be used to fit models involving first- and higher order factors simultaneously to data in a confirmatory factor analysis.

Direct expression of higher order factors

It may be useful in interpreting higher order factors to express directly their relationship to the original variables (Cattell & White, see Cattell, 1978). The path diagram of Fig. 7.10 illustrates the situation. The pattern coefficients of second-order factor A would be *b, c,* and *d* for the first-order factors C, D, and E. For the original variables G, H, I, J, and K, they would be *bg, bh, ci, cj+dk,* and *dl,* respectively. Table 7-10 shows the three pattern matrices involved. P_{01} and P_{12} are the pattern matrices obtained in the first- and second-order factor analyses. As you should verify, the third matrix P_{02} can be obtained from these by the matrix multiplication $P_{01}P_{12}$. Thus, multiplication of the two factor pattern matrices will yield the factor pattern matrix directly relating second-order factors to the original variables. For a third-order analysis, $P_{01}P_{12}P_{23}$ would yield the factor pattern P_{03} relating the third-order factors to the original variables. The extension to still higher orders is straightforward.

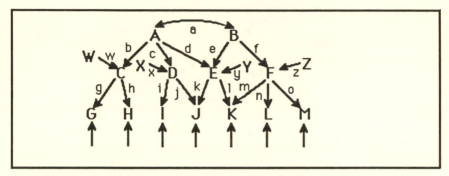

Fig. 7.10 Path diagram representing a higher-order factor analysis. C, D, E, F = 1st-order factors; A, B = 2nd-order factors; G, H, I, J, K, L, M = observed variables; W, X, Y, Z = residuals from 2nd order analysis.

Table 7-10 Factor patterns of Fig. 7.10: (a) variables related to 1st-order factors; (b) 1st-order factors related to 2nd-order factors; (c) variables related directly to 2nd-order factors

(a)					(b)				(c)		
P_{01}					P_{12}				P_{02}		
	C	D	E	F		A	B			A	B
G	g	-	-	-	C	b	-		G	bg	-
H	h	-	-	-	D	c	-		H	bh	-
I	-	i	-	-	E	d	e		I	ci	-
J	-	j	k	-	F	-	f		J	cj+dk	ek
K	-	-	l	m					K	dl	el+fm
L	-	-	-	n					L	-	fn
M	-	-	-	o					M	-	fo

Note: Dash indicates zero path. Subscripts 2, 1, 0 refer to 2nd-order factors, 1st-order factors, and variables.

Schmid-Leiman transformation

Another approach to relating higher-order factors directly to the observed variables is that due to Schmid and Leiman (1957). This representation is illustrated in Fig. 7.11 for the same case as that of Fig. 7.10. The strategy followed is a hierarchical one: the higher order factors are allowed to account for as much of the correlation among the observed variables as they can, and the lower order factors are reduced to residual factors uncorrelated with each other and with the higher order factors. Although the relative amount of

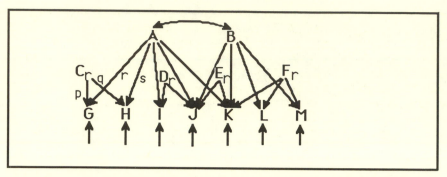

Fig. 7.11 Path diagram representing higher-order factor analysis with 1st-order factors residualized. A, B = 2nd-order factors of original analysis; C_r, D_r, E_r, F_r = residualized 1st-order factors of original analysis; G, H, I, J, K, L, M = variables.

influence attributed to the lower order factors is decreased by this, they may gain in clarity of interpretation, for each now represents the independent contribution of the factor in question. If, say, A represented a broad influence in the personality domain such as extraversion, and C, D, and E represented such component traits as sociability, impulsivity, risk taking, and the like, this procedure would allow extraversion to account for as much as possible of the intercorrelation among observable extraverted behaviors (G, H, I, J, K), and the residualized C_r, D_r, and E_r would represent effects specific to impulsivity,

risk-taking, and so on. One might well argue in particular cases whether this model better represents the underlying causal influences than does the original representation of Fig. 7.10, but when it does, the Schmid-Leiman transformation will be useful.

Basically, the procedure is as follows: The pattern matrix for the highest order (in this case the second) is obtained as for the Cattell-White transformation of the preceding section, by $P_{01}P_{12}$, based on the original first- and second-order analyses. Then, the next-lower order factors (here, the first-order factors) are residualized by scaling down their original pattern coefficients by the multiplication $P_{01}U_1$, where U_1 is a diagonal matrix of the square roots of the uniquenesses from the higher order analysis.

The reader may find this process easier to understand if it is looked at in path terms. The key step is to realize that the residualized factors C_r, D_r, etc. of Fig. 7.11 are the direct equivalents of the second-order residuals W, X, etc. in Fig. 7.10. The square roots of the uniquenesses, *u*, used in the rescaling are just the values of the paths *w*, *x*, etc. in Fig. 7.10, and the paths *p*, *q*, etc. in Fig. 7.11 are equivalent to the compound paths *wg* and *wh* in Fig. 7.10. In the Schmid-Leiman transformation we break the causation of, say, variable G into two independent paths (plus a residual). These two paths are labeled *p* and *r* in Fig. 7.11, and they correspond to *wg* and *bg* in Fig. 7.10. W is of course a causal source independent of A, by its definition as a residual.

Fig. 7.12 and Table 7-11 illustrate the procedure numerically with a simple example. (This particular example is too small actually to have yielded a determinate solution in an exploratory second-order analysis, but will suffice to illustrate the Schmid-Leiman procedure.) Fig. 7.12 (a) is the original two-level path model, whereas diagram (b) is the alternative Schmid-Leiman representation, after B and C have been transformed to uncorrelated residual factors. Diagrams (a) and (b) imply exactly the same correlation matrix and

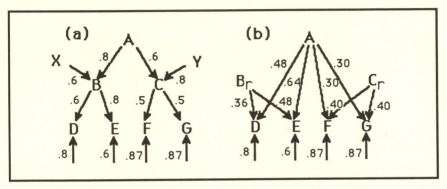

Fig. 7.12 (a) A simple two-level factor model, and (b) the same model after a Schmid-Leiman transformation. B_r and C_r are residualized versions of B and C.

communalities: for example, r_{DE} is .6 x .8 = .48 on the left, and .36 x .48 + .48 x .64 = .48 on the right. The communality of D is $.6^2$ = .36 on the left, and $.36^2 + .48^2$ = .36 on the right.

Table 7-11 shows the correlation matrix **R** among the observed variables, the correlation matrix **F** among the first-order factors that is the basis of the second-order analysis, and (left, below) the pattern matrices P_{01} and P_{12} of the first- and second-order analyses (these matrices contain the paths at the first and second levels of Fig. 7.12a). At the bottom center is the pattern matrix P_{02} relating the second-order factor A to the original variables (obtained via the Cattell-White formula), and to its right the pattern matrix P_{01} for the residualized factors B_r and C_r after the Schmid-Leiman transformation. These are proportional to the original P_{01} coefficients (bottom, left). The rescaling is by u, the square roots of the uniquenesses of the original first-order factors (top right). Note that because C_r and B_r are independent of each other and of A, the communality for each variable can be obtained by summing the squares of the coefficients across rows of P_{02} and P_{01} (lower right in the table).

Table 7-11. Matrices for Schmid-Leiman example of Fig. 7.12

R

	D	E	F	G	h^2
D	1.000	.480	.144	.144	.36
E		1.000	.192	.192	.64
F			1.000	.250	.25
G				1.000	.25

F

	B	C	h^2	u^2	u
B	1.00	.48	.64	.36	.60
C		1.00	.36	.64	.80

P_{01}

	B	C
D	.60	.00
E	.80	.00
F	.00	.50
G	.00	.50

P_{12}

	A
B	.80
C	.60

P_{02}

	A
D	.48
E	.64
F	.30
G	.30

P_{01}

	B_r	C_r	h^2
D	.36	.00	.36
E	.48	.00	.64
F	.00	.40	.25
G	.00	.40	.25

Note: P_{01} and P_{12} presumed to be obtained in a first- and second-order factor analysis of correlation matrix **R**, with the second-order analysis based on factor correlation matrix **F** from the first-order analysis. P_{02} obtained as $P_{01}P_{12}$, the Cattell-White formula. The P_{01} matrix on the right for the Schmid-Leiman residualized factors B_r and C_r is obtained as $P_{01}U_1$, where U_1 is a diagonal matrix of u, the square roots of the uniquenesses of B and C, based on the second-order analysis (upper right).

Note also that the reduced correlation matrix R_r (with communalities in the diagonal) can be obtained by adding together two independent components: $P_{02}P_{02}{'}$ and $P_{01}P_{01}{'}$, representing, respectively, the contribution of the general factor A and the group factors B_r and C_r.

Finally, the reader should note the complete equivalence of the matrix results and the path results: i.e., in this simple case the values in Fig. 7.12 (b) can be obtained virtually by inspection from Fig. 7.12 (a), as the products of the two levels of paths.

Hierarchical trait theories have long been popular, especially with British writers, since Spearman's successors first introduced a layer of group factors between *g* and the specifics. Higher order factor analysis in general, and the Schmid-Leiman transformation in particular, represent convenient ways of formalizing theories of this character.

Modes of Latent Variable Analysis

In this section we look at a number of different modes in which data can be approached in factor analysis and other latent variable methods.

R, Q, P, O, T, and S techniques of factor analysis

Cattell (1952) has suggested that sets of data to be factor analyzed may be classified along three dimensions; by considering these in pairs he defines what he calls R, Q, P, O, T, and S techniques of factor analysis. The relationships among these are summarized in Table 7-12, which is based on a similar table in Gorsuch (1983, p. 312). The reader interested in more details than are given here, or further references to the literature on this topic, will find Gorsuch a useful source.

Table 7-12 Relationships among R, Q, P, O, T, and S techniques

	What is factored	Correlation across	Example
One occasion			
R technique	measures	persons	basic personality traits
Q technique	persons	measures	personality typology
One person			
P technique	measures	occasions	individual personality structure
O technique	occasions	measures	individual psychological environment
One measure			
T technique	occasions	persons	anxiety-arousing situations
S technique	persons	occasions	anxious person types

In this view, the three basic dimensions of data are tests or measures, persons or objects measured, and situations or occasions of measurement. In the commonest form of factor analysis, one factors the relationships among tests or measures that are correlated for a sample of persons based on a single occasion of measurement. Cattell calls this *R technique*. A data matrix for typical R technique analysis is shown at the left in Table 7-13. Each of the seven tests has been given to a number of persons. A correlation coefficient is calculated for each pair of tests, i.e., between each pair of columns in Table 7-13a. The resulting 7 x 7 correlation matrix among tests is the basis for the factor analysis.

The data matrix on the right, Table 7-13b, is for the complement of R technique, *Q technique.* The form of data matrix is the transpose of that used in

209

Table 7-13 Data matrices for R and Q techniques

	(a) R technique						
	T1	T2	T3	T4	T5	T6	T7
Al	5	1	2	6	3	5	7
Ben	2	6	7	1	8	5	2
Carl	7	4	3	6	4	4	8
⋮	⋮	⋮	⋮	⋮	⋮	⋮	⋮
Zach	1	4	5	2	5	6	4

	(b) Q technique				
	Al	Ben	Carl	...	Zach
T1	5	2	7	...	1
T2	1	6	4	...	4
T3	2	7	3	...	5
⋮	⋮	⋮	⋮	⋮	⋮
T7	7	2	8	...	4

R technique--the rows are tests and the columns are people. The intercorrelations, still calculated among all possible pairs of columns and still for data gathered on a single occasion, are correlations among people rather than correlations among tests. They express how much Al is like Ben, or Ben is like Zach, on these tests. This is, you recall, the typical approach to data in multidimensional scaling; however, the measurement of resemblance in MDS is not necessarily in the form of correlations over a set of tests. In the particular example in Table 7-13b, Al and Ben are negatively correlated. They are systematically unlike each other--on those tests where Ben has relatively high scores Al scores low, and vice versa. Al and Carl, on the other hand, are positively correlated, agreeing on their high and low tests. Note that resemblances based on correlations ignore possible differences in means. Al and Carl's correlation reflects the fact that they show the same *pattern* of scores, even though Carl's scores tend to be systematically higher. The correlation between them would not change if we were to add two points to every one of Al's tests, although this would make Al and Carl's scores more alike in absolute terms. Nor would the correlation decrease if we were to add 10 points to each one of Carl's scores, although in some ways this would make Al and Carl very different. For these reasons one might prefer sometimes to use another measure of association than an ordinary Pearson correlation for Q technique--for example, some measure of distances between profiles (see Overall & Klett, 1972, Chapter 8, for a discussion).

One should also be aware that Q technique correlations can be quite sensitive to the scales of the tests over which they are computed. Merely changing the scoring of test 3 in Table 7-13, to give 10 points per item instead of one, although trivial in concept, will in fact drastically affect the correlations-- for example, it changes the correlation over the 7 tests between Al and Ben from -.81 to +.92. (Can you see why?) For this reason, it is often desirable to standardize the rows of the data matrix (i.e., express the test scores in standard score form) prior to doing the correlations for a Q-type factor analysis-- particularly if the test scores are in noncomparable units. (This is often referred to as *double-centering* the data matrix, because the correlation itself effectively standardizes by columns.)

R technique seeks the dimensions underlying groupings of tests or measures and might be used, for example, in a study of basic personality traits. Q technique seeks the dimensions underlying clusters of persons and might be used, say, in a study of personality types. As we noted in a related context in discussing multidimensional scaling, the two approaches might in fact lead to the same underlying latent variables: Either a study of personality scales or of person types might lead one to an introversion-extraversion dimension. Nevertheless, the routes taken and the intermediate products of the analyses would be quite different in the two cases.

The next two techniques, P and O, are also complementary to one another, but they both use just a single subject, tested on repeated occasions. In *P technique*, shown on the left in Table 7-14, one considers two measures similar if scores on them tend to vary together over occasions in the life of an

Table 7-14 Data matrices for P and O techniques

(a) P technique						(b) O technique				
	T1	T2	T3	...	Tm		Day1	Day2	Day3	... DayN
Day1	7	1	2	...	8	T1	7	4	2	... 1
Day2	4	2	3	...	5	T2	1	2	6	... 7
Day3	2	6	8	...	3	T3	2	3	8	... 6
:	:	:	:		:	:	:	:	:	:
DayN	1	7	6	...	2	Tm	8	5	3	... 2

individual. In the table, measures T1 and Tm appear to go together, as do T2 and T3, with the two sets tending to be negatively related. P technique is best suited for use with measures of states, such as moods or motive arousal, which can be expected to vary from day to day in the life of an individual. Its merit is that it can give a picture of the mood or motive structure of that particular person. Some personality psychologists, who have objected to the usual R-type factor analysis as only yielding a picture of a mythical "average person" (no example of which may actually exist!), should find a P technique approach more congenial.

An illustration of a study using P technique is that of Cattell and Cross (1952), in which multiple measures designed to assess the strengths of a number of motivational states (anxiety, self-confidence, sex drive, fatigue, and the like) were obtained twice daily for a particular individual--a 24-year-old drama student--over a period of 40 days. A factor analysis of the intercorrelations of these measures over the 80 occasions yielded some patterns much like those that had been found in previous R-type researches, but others that appeared to be idiosyncratic to this particular individual--or at any rate, to his life during this period.

O technique, the complement of P technique, has not apparently been much used, although it is interesting in principle. Its correlations (Table 7-14b) are based on the similarity of occasions in one person's life, assessed across a multiplicity of measures. It asks which are the occasions that go together in terms of a person's reactions to them--which situations arouse anxiety, which are challenging, which depressing. One might think of this as a way of getting at the structure of a person's psychological environment, of the events to which he or she responds in characteristic ways. For the same reasons as in Q technique, preliminary standardization of scores on the measures used will often be desirable in O technique.

The final pair of complementary approaches, *T* and *S techniques*, seem also not to have been much explored. They restrict themselves to a single response measure but assess it across both persons and situations. In T technique one looks at resemblances among situations in their effect on the response measure, and in S technique at resemblances among persons. For example, in a study such as that of Endler, Hunt, and Rosenstein (1962), in which ratings of anxiety were obtained across both persons and situations, one could either factor the different types of situations to study the relationships among anxiety-arousing situations (T technique), or factor the persons to obtain a typology of persons based on the situations that arouse their anxiety (O technique). In either case one might infer latent dimensions such as physical versus social anxiety, or realistic versus imaginary fears.

Three-mode factor analysis

The six types of factor analysis described in the preceding section can be considered to represent different ways of collapsing a three-dimensional rectangular data matrix--see Fig. 7.13. The two horizontal dimensions of the cube are measures, and situations or occasions, the vertical dimension is persons. Any point within the cube represents the score of a particular person on a particular measure on a particular occasion.

If we take a slice off the left front face of the cube (or any slice parallel to it), we have data from a single occasion and hence an R or a Q technique study, depending on whether we choose to run our correlations vertically or horizontally. If we take a slice off the right face, or parallel to it, we have data from a single measure, and hence a T or S technique study. And if we take a slice off the top, or any horizontal slice, we have data from a single person, and P or O technique, depending on whether we calculate our correlations among the measures or the occasions.

But can we do a single, overall analysis of the whole data cube? Yes, we can. The procedure is known as three-mode factor analysis and was developed by Ledyard Tucker (1964). We do not attempt to describe the

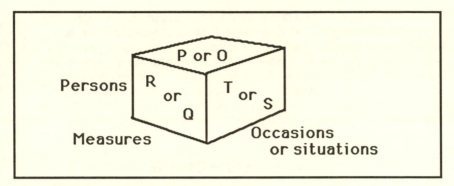

Fig. 7.13 A three-dimensional data matrix.

method at the level of calculational detail, but, roughly, it results in three sets of factors resulting from the analysis of correlations involving measures, persons, and situations, and a *core matrix* that relates the three separate sets of factors.

Tucker presents an analysis of the Endler, Hunt, and Rosenstein anxiety data. The measures were different reponses that might be associated with anxiety (e.g., "heart beats faster," "need to urinate frequently"); the situations were such potentially anxiety-arousing situations as making a speech or going on an initial date; and the persons were the student subjects filling out the ratings. Factors were reported for each of the three modes separately: for example, "heart beats faster" and "get uneasy feeling" went together on a measures factor; "speak before large group" and "job interview" went together on a situations factor; and three person dimensions emerged among the subjects. The core matrix showed relationships involving all three modes: for example, one of the person types showed various distress responses in interpersonal situations, while two other types tended to show exhilaration; the latter two types differed, however, in their responses to situations involving inanimate or unknown dangers.

Three-mode analyses in MDS and structural equation analysis

Analogous approaches to data in three modes occur in multidimensional scaling and structural equation analysis, although they have not been so formally systematized as in factor analysis. Individual-differences MDS is a type of three-mode analysis, in which a two-dimensional matrix of object resemblances is extended to a third mode of persons. A multitrait-multimethod matrix is three mode: traits, methods, and persons. And so is a structural equation analysis of events over time: Several measures on each of a sample of persons are taken on each of several occasions. In each case the simultaneous analysis over the three modes is in principle capable of providing information unattainable from any two modes considered separately.

Many opportunities have yet to be explored for extending such analyses to new kinds of problems. You might want to think about what it might mean to do, say structural equation or MDS analyses of O or S types in your own substantive area of interest. Also, there is no law that says that three modes is the limit. For example, Cattell in one place discusses as many as 10 modes (1966b). Finally, there are many variations possible *within* any single design-- for example, instead of achievement test scores on schoolchildren across grades, how about economic indicators on countries across decades? The risk of ever having to say "everything has been done" seems negligible.

In Closing, Some Caveats

In a thoughtful article entitled "Some cautions concerning the application of causal modeling methods," Cliff (1983) gives some gentle warnings and sensible advice to the users of programs such as LISREL: "Initially, these methods seemed to be a great boon to social science research, but there is some danger that they may instead become a disaster, a disaster because they seem to encourage one to suspend his normal critical faculties. Somehow the use of one of these computer procedures lends an air of unchallengeable sanctity to conclusions that would otherwise be subjected to the most intense scrutiny" (p. 116).

I hope that if you have got this far in this book you have enough sense of how these models work and do not work, and of some of the vicissitudes to which they are subject, that "unchallengeable sanctity" will not characterize your attitude toward conclusions drawn from their use. But it is worth reminding ourselves briefly of the four principles of elementary scientific inference that Cliff suggests are particularly likely to be violated in the initial flush of enthusiasm of causal modelers:

> The first principle is that the data do not confirm a model, they only fail to disconfirm it, together with the corollary that when the data do not disconfirm a model, there are many other models that are not disconfirmed either. The second principle is that *post hoc* does not imply *propter hoc.* That is, if *a* and *b* are related, and *a* followed *b* in time, it is not necessarily true that *b* caused *a*. The third principle is that just because we name something does not mean that we understand it, or even that we have named it correctly. And the fourth principle is that *ex post facto* explantions are untrustworthy. (pp. 116-117)

Let us look at each of these four principles in a little more detail.

The unanalyzed variable

Suppose that a particular model "fits"--i.e., it is not rejected, given the data. That does not mean that other models would not fit equally well, maybe even better. Suppose, for example, there is some variable V that we have overlooked that is related to the variables X, Y, and Z, which we have included in our model. We can run LISREL forward, backward, and upside down on a model of X, Y, and Z, but but it will not tell us that a model with V in it would have fit better. We can only be reasonably sure that if there is such a V and it has causal effects on variables included in our model, our estimates of some of the paths in our model will be wrong. As Cliff puts it: "These programs are not magic. They cannot tell the user about what is not there" (p. 118).

Post hoc is not *propter hoc*

Cliff cites an incautious author who concludes, on the basis of a significant arrow in a path diagram, that "Father's Occupation caused [Child's] Intelligence." Cliff goes on, "It may be that it does, but somehow I doubt it. It seems unlikely that, if ever the causal variables involved in scores on modern 'intelligence' tests are sorted out, one's father's occupation will ever be one of them" (p. 120). Of course, there may be variables *correlated* with father's occupation that *do* play a causal role, but that takes us back to the preceding point.

To make things worse, time of measurement is not always a safe guide to the sequence of events. "Consider the possibility that we measure a child's intelligence in the fifth grade and her father's occupation when she is in the tenth" (p. 120). Should we then put in a causal arrow leading from the earlier to the later event?

The fact that we can name it does not mean we know what it is

Latent variables are only defined by way of their associations with manifest variables. Because we are always to some degree wrong about what our manifest variables mean (there is always some degree of invalidity and unreliability of measurement), Cliff says, ". . . we can only interpret our results very cautiously unless or until we have included enough indicators of a variable in our analysis, and have satisfied not only ourselves but skeptical colleagues and critics that we have done so" (p. 121). Even a "confirmatory" factor analysis does not escape these problems. It just tells us that we have *one* set of parameters that is consistent with the data. "There are typically an infinity of alternative sets of parameters which are equally consistent with the data, many of which would lead to entirely different conclusions concerning the nature of the latent variables" (pp. 122-123).

Ex post facto explanations are untrustworthy

Once a model has been modified to make it fit better to a given data set, one can no longer take the probability values associated with subsequent goodness-of-fit tests at face value. If a model has been adjusted on the basis of its fit or lack of fit to a particular body of data, its statistical status is precarious until it can be tested on a new body of data that did not contribute to the adjustment.

One way to deal with this problem is cross-validation. Split the initial data set in half, play around with model-fitting on one half of the data until you get a model you are happy with, and *then* carry out the statistical test--once--on the unused half of the data. The χ^2 will then be legitimate. This procedure has its disadvantages--for one thing, it requires twice the sample size--but it has the preeminent advantage of not leaving the investigator and his readers with results which "they know are unstable to an unknown degree" (p. 124).

Let me add a couple of additional caveats to Cliff's four.

It's always "If . . . ."

Statistical tests in maximum likelihood model fitting are always of the form "*If* such-and-such a model is assumed (and the other statistical requirements are met), we can conclude . . . ." That first "if" can be a big one. In fact, frequently you can make very drastic changes in the meaning of a path diagram--for example, by reversing the direction of one or more of its causal arrows--and still have a model that fits the data exactly as well as before, but with quite different values for its paths. (Stetzl, 1986, discusses the conditions under which this will happen.) Let us consider the example shown in Table 7-15 and Fig. 7.14.

The six variables D through I represent two measures for each of three latent variables: a mother's verbal aptitude, her child's verbal aptitude, and the amount of reading aloud to the child that the mother has done.

Table 7-15 Correlations among six hypothetical variables (for example of Figure 7-14), N = 100

	D	E	F	G	H	I
D	1.00	.56	.38	.34	.50	.50
E		1.00	.43	.38	.58	.58
F			1.00	.72	.43	.43
G				1.00	.38	.38
H					1.00	.64
I						1.00

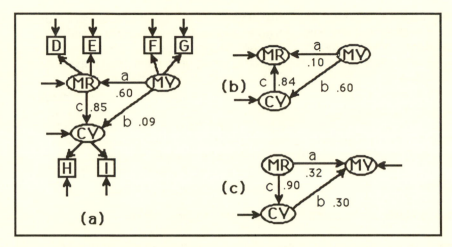

Fig. 7.14 Three alternative models of relations among MV, mother's verbal aptitude, CV, child's verbal aptitude, and MR, mother's reading to child. (Same measurement model throughout.)

On the left in the figure, model (a), is a model whose structural portion assumes that a mother's reading aloud to her child has a causal effect on the development of the child's verbal aptitude. The model also assumes that a mother's verbal aptitude may affect how much she is inclined to read to the child, as well as possibly having a direct effect on the child's verbal aptitude, an effect that might, for example, occur via the genes. If we solve this model for the correlations in Table 7-15, we obtain a very good fit indeed, the χ^2 is .02, based on 7 df. The value of the path from mother's verbal aptitude to her reading to the child (path a) is .60, the direct path from her own verbal aptitude to her child's verbal aptitude (path b) is .09, and the effect of mother's reading on child's verbal aptitude (path c) is .85. Obviously, mother's reading aloud is a key variable in explaining the child's verbal aptitude.

But suppose that we were to entertain a quite different hypothesis, namely, that a mother's reading to a child has no effect whatever on its verbal aptitude; but, on the other hand, that the amount of reading a mother does to her child over the years is affected by how much the child enjoys it, which is in part a function of the child's verbal aptitude. We now have the structural model shown as (b). If we fit that model to the data, we obtain exactly the same good fit, and the same low χ^2. But the values of the paths, and the interpretation of the model, are now quite different. There is a substantial direct determination of child's verbal aptitude by mother's verbal aptitude ($b = .60$). Mother's verbal aptitude has only a very minor effect on how much she reads to the child ($a = .10$). Path c remains strong (.84), although now, of course, it represents an entirely different causal effect. A developmental psychologist with hereditarian thoeretical preferences might like this second model better, with its direct

217

mother-child transmission, and a psychologist of environmentalist inclinations might fancy the first. The point, however, is that both models are exactly equivalent in their fit to the data. And so, for that matter, are others--including such unlikely models as (c)--which both psychologists might be pained to find has an identical χ^2 of .02, for a = .32, b = .30, and c = .90.

The practical moral: Think about each causal arrow in your path diagram. If there is reasonable doubt about which direction it goes, it might make sense to try solutions both ways and see how much difference this makes for other paths in the model. Sometimes the effects of such a change will be quite localized. If so, the interpretation of other paths may still be secure. If not . . . . Well, surely you would want at least to be aware of this fact when discussing your results.

Large samples are essential for exploratory model fitting

As we have seen, many practioners of model fitting will modify their initial model, if it happens to fit poorly, by adding and deleting paths in an effort to secure a more adequate fit to the data. A simple piece of advice: Do not try this with small samples, unless you are very brave indeed.

In a recent study, MaCallum (1986) took known models, imposed simple specification errors (for example, a path might be omitted or an extra one added), and fit the models to random samples of data from populations in which the true models held. All models had a correctly specified measurement portion--the errors occurred only in the structural model. The procedure used was a typical one for exploratory model fitting: If a model does not fit, make the single change that most improves its fit. Repeat as necessary until a nonsignificant χ^2 is achieved. Then test for and delete any unnecessary paths.

The results were moderately complex, but the following examples should give a feel for them. For a sample size of 300 cases and just a single omitted path, only 10 of 20 attempts were successful in reaching the true model. With the same sample size and a more poorly specified model (two paths omitted, one added), the true model was never achieved in 20 tries, although sometimes one or two correct steps toward it were taken. With the latter model and N = 100 there were many problems, such as improper solutions or a failure to reject the initial model. Only 8 of 20 tries even got as far as making one legitimate change, and in 7 of the 8 it was a wrong one.

The moral: Don't expect magical results from exploratory model fitting, although minor improvements may be possible. In any event, use as large samples as you can and plan to check your results on new data. It is probably foolish to take the results of exploratory model-modification seriously if your N is 100 or less--certainly in the absence of cross-validation.

In conclusion

Neither Cliff nor I would wish to discourage you from the use of causal model-fitting methods, which in his view represent "perhaps the most important and influential statistical revolution to have occurred in the social sciences" (1983, p. 115). He concludes, and I can only echo: "programs such as LISREL and its relatives provide completely unprecedented opportunities . . . . With their aid, conclusions can be made which heretofore would have been impossible, but only provided the analysis is approached intelligently, tough-mindedly, and honestly" (p. 125).
     Go do it.

Chapter 7 Notes

   Multivariate path models. For examples of doing it the hard way see Neale and Fulker (1984) and Vogler and DeFries (1985)--both are bivariate cases handled by nonmatrix methods.
   Nonlinear effects. See also Busemeyer and Jones (1983) on dealing with multiplicative effects.
   Multidimensional scaling. A paper by Denison (1982) compares MDS and structural modeling. The methods of Guttman and his colleagues are closely related to MDS, e.g., the procedure called Smallest Space Analysis. For this and a whole family of related methods, see Lingoes (1973).
   Nonlinear factor analysis. A recent paper by Etezadi-Amoli and McDonald (1983) describes a "second generation" version. See Lingoes and Guttman (1967) for an approach to "nonmetric factor analysis."
   Modes of factor analysis. Cronbach (1984) discusses R, P, Q, etc. techniques. Kroonenberg (1983) provides an extensive annotated bibliography on 3-mode factor analysis. Three-mode longitudinal applications are discussed by Kroonenberg, Lammers, and Stoop (1985).
   Statistical developments. Relaxation of the assumption of multivariate normality, especially with regard to kurtosis, is addressed by Browne (1984) and Bentler (1983a,b), leading to the more general family of elliptical distributions, and to asymptotically distribution-free methods (see also notes to Chapter 2). Other approaches avoiding normality assumptions, such as bootstrapping, are beginning to be explored for latent variable problems-- for example, Chatterjee (1984) gives an application to exploratory factor analysis (actually, in his example, component analysis). Satorra and Saris (1985) provide a way of estimating the power of the likelihood ratio test in covariance structure analysis against a specified alternative hypothesis. The effects of selectivity in nonrandom sampling are considered by Muthén and Jöreskog (1983). Muthén (1983) also discusses latent variable modeling when the data are in the form of categories rather than continuous variables. Lee (1986) discusses the handling of missing data in structural modeling.

Dong (1985) suggests a way of dealing with singular matrices. And Bentler and Lee (1983) address the question of a proper statistical approach to the structural analysis of correlation matrices (i.e., not just treating them as covariance matrices).

Critique and controversy. For other examples of criticisms of the causal modeling enterprise, see Martin (1982) and Baumrind (1983); Huba and Bentler (1982) reply to the former.

Chapter 7 Exercises

1. Table 7-13 shows means, standard deviations, and correlations on four hypothetical masculinity-femininity scales in samples of men (below diagonal) and women (above diagonal). Is it reasonable to conclude that there is a general masculinity-femininity latent variable that accounts both for the interrelationships among the measures and the differences in mean and variance between the samples? Are there differences between the sexes in how the tests measure this factor?

Table 7-13 Data for problem 7-1 (women above diagonal, men below)

Scale	A	B	C	D	M	SD
A	1.00	.48	.10	.28	5.2	.98
B	.50	1.00	.15	.40	7.0	1.00
C	.12	.16	1.00	.12	6.1	.99
D	.45	.70	.17	1.00	8.3	1.03
M	7.5	10.2	7.0	11.1	N	208
SD	1.08	1.15	1.01	1.18	200	

2. (a) Write from Fig. 7.3 the path equation C_{S1D2} for the covariance of sons on trait 1 with daughters on trait 2. (b) Obtain the numerical value for that covariance and for the one for C_{S1D1} given in the text, by substituting the appropriate values of paths and correlations from Table 7-14 (assume standardized variables). (c) Set up the matrices required for the expression for C_{SD} given in the text and carry out the matrix calculation via MINITAB or some other program with matrix facilities; check for consistency with the results of part (b).

Table 7-14 Data for Problem 7-2

	Correlations				Paths			
	F1	F2	M1	M2	S1	S2	D1	D2
F1	1.00	.50	.20	.10	.40	.10	.30	.20
F2		1.00	.10	.30	.10	.20	.20	.40
M1			1.00	.60	.50	.20	.50	.30
M2				1.00	.20	.40	.30	.50

3. Write the path equations for V_A, $C_{B,AC}$, $C_{AC,AD}$, and V_Y, from Fig. 7.6.

4. Construct a one-dimensional model whose stress value lies between those of models #1 and #2 in Table 7-6, and another one-dimensional model with stress lower than that of model #2.

5. Apply a Varimax rotation to the factor pattern of Table 7-9. How would you interpret the resulting rotated factors in terms of the equations in Table 7-8?

6. Take the oblique Quartimin solution obtained in Chapter 6, Problem 7. Carry out a second-order factor analysis: Extract a single principal factor from the **F** obtained in that analysis. Report P_{12}.

7. Relate the second-order factor in the preceding problem directly to the original first-order variables.

8. Subject the analysis of Problems 6 and 7 to a Schmid-Leiman transformation. Compare the final communalities to those from the original principal factors.

9. Carry out an exploratory Q-technique factor analysis of the data in Table 7-13, using the scores for the 4 persons on all 7 tests. Comment on your results.

10. Suggest an example of T technique that might be relevant to experimental psychology.

11. Is the study of Judd and Milburn (see Chapter 4) a four-mode structural analysis? (Give your reasoning.)

12. Think of an additional caveat that you might offer a beginner in causal modeling.

Appendix A: Simple Matrix Operations

This appendix reviews some basic aspects of matrix algebra, confining itself to those that are used in this book and proceeding largely by example rather than by formal definition and proof.

Matrices

A *matrix* is a rectangular array of numbers. Examples of three matrices, **A**, **B**, and **D**, are given in Fig. A.1. Matrix **A** has *dimensions* 4 x 3 (the number of *rows* precedes the number of *columns*). **B** and **D** are *square* 3 x 3 matrices; they can alternatively be described as being of *order* 3. Matrices **B** and **D** are also *symmetric* matrices: Each row of the matrix is identical with the corresponding column, so that the matrix is symmetrical around the *principal diagonal* that runs from its upper left to lower right. A symmetric matrix is necessarily square, but a square matrix is not necessarily symmetric: The first three rows of matrix **A** would constitute a square, nonsymmetric matrix.

1	4	7	1.00	.32	.64	2	0	0
6	2	5	.32	1.00	.27	0	-3	0
3	3	0	.64	.27	1.00	0	0	-1
4	6	1						
	A			**B**			**D**	

Fig. A.1 Some examples of matrices.

Matrix **B** happens to be a familiar example of a symmetric matrix, a correlation matrix. A variance-covariance matrix would provide another example. Matrix **D** is a *diagonal* matrix: all zeroes except for the principal diagonal. If the values in the diagonal of **D** were all 1's, it would have a special name: an *identity* matrix, symbolized **I**. A matrix of all zeroes is called a *null* matrix.

Matrices are ordinarily designated by bold-face capital letters.

The transpose of a matrix

The *transpose* of a matrix is obtained by interchanging each row with its corresponding column. Thus, the transpose of matrix **A** in Fig. A.1, conventionally designated **A´**, is:

$$\begin{matrix} 1 & 6 & 3 & 4 \\ 4 & 2 & 3 & 6 \\ 7 & 5 & 0 & 1 \end{matrix}$$

in which the first column of **A** becomes the first row of **A´**, and so on. In the case of a symmetric matrix, **A´ = A**, as you can see if you attempt to transpose matrix **B** or **D** in Fig. A.1.

Note that the transpose of a 4 x 3 matrix will have the dimensions 3 x 4 because rows and columns are interchanged. The transpose of a square matrix will be another square matrix of the same order, but it will not be the same matrix unless the original matrix was symmetric.

Vectors and scalars

A single row or column of numbers is called a *vector*. Examples of *column* and *row* vectors are shown in Fig. A.2. The two vectors are of *length* 4 and 3, respectively. Column vectors are designated by lower case bold-face letters-- e.g., vector **a**. Row vectors, as transposes of column vectors, are marked with a prime symbol--e.g., vector **b´**.

```
      a:  1        b´:  .3  1.0  .5
          2
          4        c:  17.3
          3
```

Fig. A.2 Column and row vectors, and a scalar.

Single numbers, of the kind familiar in ordinary arithmetic, are referred to in matrix terminology as *scalars* and are usually designated by lower case roman letters--e.g., scalar c in Fig. A.2.

Addition and subtraction of matrices

Two matrices, which must be of the same dimensions, may be *added* together by adding algebraically the corresponding elements of the two matrices, as shown for the two 3 x 2 matrices **A** and **B** in Fig. A.3. A matrix may be *subtracted* from another by reversing the signs of its elements and then adding, as in ordinary algebra. Examples are shown at the right in Fig. A.3.

```
   1  2     0  1      1  3     1  1     -1 -1
   4  6     0 -1      4  5     4  7     -4 -7
   5 -2     3 -1      8 -3     2 -1     -2  1

     A         B      A+B      A-B      B-A
```

Fig. A.3 Matrix addition and subtraction.

Two column vectors of the same length, or two row vectors of the same length, may be added or subtracted in the same way. (As, of course, may two scalars.)

Multiplication of vectors and matrices

A row vector and a column vector of the same length may be multiplied by obtaining the sum of the products of their corresponding elements, as illustrated in Fig. A.4 for **a´b**. Note that this is a multiplication in the order *row* vector times *column* vector. In matrix arithmetic the order of matrices or vectors in a

$$
\begin{array}{llll}
\mathbf{a´}: 1\ 0\ 2\ 3 \quad \mathbf{b}: & 1 & 1 \times 1 = 1 \\
& 2 & 0 \times 2 = 0 \\
& 4 & 2 \times 4 = 8 \\
& 3 & 3 \times 3 = \underline{9} \\
& & \mathbf{a´b} = 18
\end{array}
$$

Fig. A.4 An example of the vector multiplication **a´b**.

multiplication is *not* the indifferent matter that it is in scalar arithmetic, where ab = ba. The product **ba´** is something entirely different from **a´b** (as we see shortly). Two matrices may be multiplied by multiplying each of the row vectors of the first matrix in turn times each of the column vectors of the second matrix. Each of these vector multiplications yields a single number that constitutes an element of a row of the product matrix. A step-by-step example is given in Fig. A.5. Notice that the result matrix has as many *rows* as the *first* matrix, and as many *columns* as the *second*, and that for multiplication to be possible, the rows of the first matrix and the columns of the second must be equal in length. Two matrices are said to *conform* for multiplication when this last condition holds--an easy way of checking is to see that the second dimension of the first matrix agrees with the first dimension of the second: that is, a 3 x 5 matrix can be multiplied times a 5 x 2, or a 7 x 2 times a 2 x 3. The middle numbers in the sequence must match; the outer numbers give the dimensions of the result. In the first example, the 5's match and the result will be a 3 x 2 matrix; in the second example, the 2's match and the result will be 7 x 3. In Fig. A.5, the multiplication was of a 3 x 2 times a 2 x 2 matrix; the middle 2's match and the result was 3 x 2. Working through a sample case or two will make it evident to you why these rules hold.

These principles generalize to longer series of matrix multiplications: If **W**, **X**, **Y**, and **Z** are, respectively, of dimensions 4 x 2, 2 x 3, 3 x 7, and 7 x 5, the multiplication **WXYZ** can be carried out and the result will be of dimension 4 x 5. (You can see this by carrying out the steps successively: 4 x 2 times

```
1 3    4 5        1st row of  A,  1st col of B      1 x 4 = 4
2 1    2 2                                          3 x 2 = 6
0 2                                                       10
 A      B          1st row of A,   2nd col of B      1 x 5 = 5
                                                     3 x 2 = 6
                                                          11
                   2nd row of A,   1st col of B      2 x 4 = 8
                                                     1 x 2 = 2
                                                          10
                   2nd row of A,   2nd col of B      2 x 5 =10
                                                     1 x 2 = 2
                                                          12
                   3rd row of  A,  1st col of B      0 x 4 =  0
                                                     2 x 2 =  4
                                                             4
                   3rd row of  A,  2nd col of B      0 x 5 = 0
                                                     2 x 2 = 4
result AB:  10  11                                          4
            10  12
             4   4
```

Fig. A.5 Step-by-step example of the matrix multiplication **A** times **B**.

2 x 3 is proper and will yield a 4 x 3 matrix; 4 x 3 times 3 x 7 will work and yield a 4 x 7 matrix; and so on.) The rules also hold for vectors, considered as 1 x n or n x 1 matrices. Thus, a 1 x 4 row vector times a 4 x 1 column vector yields a 1 x 1 single number result (as we have seen in Fig. A.4). A 4 x 1 column vector times a 1 x 4 row vector, on the other hand, would produce a 4 x 4 matrix as an answer, and that matrix would be obtained by applying the regular rules of matrix multiplication: taking each row (of length 1) of the column vector and multiplying it successively by each column of the row vector to yield the elements of the rows of the result vector. (You might want to verify that for the two vectors of Fig. A.4, the first two rows of the product **ab**′ would be 1 2 4 3 and 0 0 0 0, and the first two rows of **ba**′ would be 1 0 2 3 and 2 0 4 6.)

Because the order of matrix multiplication is important, the terms *pre* and *postmultiplication* are often used to eliminate ambiguity. In the product **AB**, **B** is said to be premultiplied by **A**, or **A** to be postmultiplied by **B**. In the product **a**′**Ba**, the matrix is pre and postmultiplied by the vector. (Incidentally, can you see that **B** must be a square matrix, that the result will be a scalar, and why one would seldom run across the alternative product **aBa**′?)

Some special cases of matrix multiplication

The basis of these rules will be self-evident if you work through an example or two.

(1) A matrix pre or postmultiplied by a null matrix yields a null matrix. (The null matrix acts like a zero in scalar arithmetic.)

(2) A matrix pre or postmultiplied by an identity matrix is unchanged. (The identity matrix acts like a 1 in scalar arithmetic.)

(3) Premultiplying a matrix by a diagonal matrix rescales the rows of the matrix by the corresponding elements of the diagonal matrix; postmultiplying by a diagonal matrix rescales the columns. (Try **AD** or **DB** in Fig. A.1, for example.)

(4) Pre or postmultiplying a matrix by its transpose can always be done and yields a symmetric matrix.

Multiplying a vector or matrix by a scalar

Multiplying a matrix or vector by a scalar is done by multiplying every element in the matrix or vector by that scalar. In a series of matrix operations, the location of a scalar does not matter, and may be changed at will: **ka´BC = a´kBC = a´BkC = a´BCk**, where k is a scalar. (But of course **ka´BC** doesn't equal **kBa´C** or **ka´CB**--the vectors and matrices cannot in general be reordered.)

The inverse of a matrix

There is no operation of matrix division as such, but a matrix *inverse* is the matrix analogue of a reciprocal of a number in scalar arithmetic, so multiplying by an inverse is the matrix equivalent of dividing.

The inverse of a matrix **A**, symbolized by **A**$^{-1}$, is a matrix such that **AA**$^{-1}$ or **A**$^{-1}$**A** equals an identity matrix. (Just as k x 1/k = 1.) Only square matrices have inverses, and not all of them do--if a matrix has some rows or columns that are linearly predictable from others, it will have no inverse. A matrix that has no inverse is called *singular*. Matrix inversion is a basic step in solving matrix equations: If **BX = A** and **B** has an inverse, you can solve for **X** by premultiplying both sides of the equation by **B**$^{-1}$, i.e.:

$$\textbf{BX = A}$$
$$\textbf{B}^{-1}\textbf{BX = B}^{-1}\textbf{A}$$
$$\textbf{IX = B}^{-1}\textbf{A}$$
$$\textbf{X = B}^{-1}\textbf{A}$$

Obtaining the inverse of a matrix tends in general to be a large computational task. Let a computer do it. You can always check to see whether the result it has given you is correct by carrying out the multiplication

AA^{-1}, which should equal I within rounding error.

Some useful facts about inverses:

(1) The inverse of the transpose of a matrix is the transpose of its inverse: $(A')^{-1} = (A^{-1})'$.

(2) Taking the inverse of an inverse yields the original matrix: $(A^{-1})^{-1} = A$.

(3) The inverse of a symmetric matrix is also symmetric.

```
1 0 0      1 0 0      2 0  0     1/2  0    0
0 1 0      0 1 0      0 3  0      0  1/3   0
0 0 1      0 0 1      0 0 -4      0   0  -1/4

  I          I⁻¹         D            D⁻¹


4 2            1    ⎡ 3 -2⎤    1.5  -1.0
5 3         ─────── ⎢-5  4⎥  = -2.5   2.0
            12 - 10 ⎣     ⎦
  A                               A⁻¹
```

Fig. A.6 Some special cases of matrix inversion.

A few special cases of matrix inversion that do not require extensive computation are illustrated in Fig. A.6. You might want to verify these by showing that $AA^{-1} = I$ in each case.

(1) The inverse of an identity matrix is itself.

(2) A diagonal matrix can be inverted by replacing each diagonal element by its reciprocal.

(3) The inverse of a 2 x 2 matrix, for example:

$$\begin{matrix} a & b \\ c & d \end{matrix}$$

may be obtained by interchanging the two diagonal elements a and d, changing the signs of the two off-diagonal elements b and c, and multiplying the result by the scalar $1/(ad - bc)$.

Inverse or transpose of a product

The *transpose* of a product of matrices is equal to the product of the transposes of the matrices, *taken in reverse order:* $(ABCD)' = D'C'B'A'$.

The *inverse* of a product of matrices is equal to the product of the inverses of the matrices, taken in reverse order:

$$(ABCD)^{-1} = D^{-1}C^{-1}B^{-1}A^{-1}$$

Of course, the latter case presupposes that the necessary inverses exist, i.e., that **A**, **B**, **C**, and **D** are square, nonsingular matrices (they must also all be of the same order for the multiplication to be possible).

Eigenvalues and eigenvectors of a correlation or covariance matrix

If **C** is an $m \times m$ variance-covariance or correlation matrix, it can be decomposed into a matrix product $\mathbf{VL^2V'}$, where **V** is a square $m \times m$ matrix whose columns are called the *eigenvectors* of matrix **C**, and $\mathbf{L^2}$ is a diagonal matrix of numbers, customarily arranged in descending order of size, called the *eigenvalues* of **C**. **V** and $\mathbf{L^2}$ are so chosen as to have the following additional properties:

$$\mathbf{V'V = I},$$
$$\text{and} \quad \mathbf{Cv} = l^2\mathbf{v},$$

where l^2 is any one of the eigenvalues and **v** the corresponding eigenvector; that is, the eigenvectors are mutually *orthogonal* (= uncorrelated), and each has the property that when it postmultiplies the matrix **C** the result is a vector proportional to itself, the coefficient of proportionality being the eigenvalue.

Eigenvalues and eigenvectors are also sometimes known as characteristic roots and characteristic vectors, or latent roots and latent vectors.

The calculation of eigenvalues and eigenvectors, even more than the calculation of inverses, is a task for computers. Again, you can always verify that the results the computer gives you have the properties specified. Fig. A.7 gives an example of eigenvectors and eigenvalues for the correlation matrix of Fig. A.1.

1.00	.32	.64	1.846	0	0	−.641	.261	.721
.32	1.00	.27	0	.797	0	−.443	−.894	−.070
.64	.27	1.00	0	0	.358	−.627	.364	−.689
	C			$\mathbf{L^2}$			**V**	

$$\text{tr } \mathbf{L^2} = 3.00 \quad |\mathbf{C}| = .526$$

Fig. A.7 Eigenvalues and eigenvectors of a correlation matrix.

A few useful additional properties of eigenvalues:

(1) The sum of the diagonals of **C** and $\mathbf{L^2}$ are the same; i.e., the sum of the eigenvalues of a matrix equals the trace of the matrix.

(2) The product of the eigenvalues of a matrix is called the *determinant* of the matrix. The determinant of matrix **C** is symbolized $|\mathbf{C}|$. A singular matrix, one which has one or more rows or columns linearly predictable from others, has one or more zero eigenvalues and thus a determinant of zero.

(3) The number of nonzero eigenvalues of a matrix is called the *rank* of the matrix, which is the number of nonredundant rows or columns it contains.

Appendix B: IPSOL

The simple program for iterative path solutions described in the text is given below. It is written in the FORTRAN computer language, which can be used on many computers. Lines beginning with an asterisk are comment lines, in ordinary English, not FORTRAN. If you are unfamiliar with FORTRAN, the comment lines should still give you a general sense of how the program is proceeding about its business.

```
          PROGRAM IPSOL(INPUT,OUTPUT)
**A SIMPLE PROGRAM FOR ITERATIVE PATH SOLUTIONS.
**EXECUTIVE PROGRAM--READS INPUT, CALLS MINIMIZATION ROUTINE,
*  PRINTS OUTPUT.  DIMENSIONED FOR UP TO 200 OBSERVED VALUES
*  AND 20 UNKNOWN PARAMETERS TO BE SOLVED FOR.
**INPUT: (1) TITLE--ONE LINE OF UP TO 80 CHARACTERS.
*          (2) NOV=NO.OF OBSERVED VALUES; NU=NO.OF UNKNOWN
*              PARAMETERS;NSOL=NO.OF SOLUTIONS; ISTEP=INITIAL STEP
*              VALUE (IF LEFT BLANK, USES .1); INTEGERS, IN 3-DIGIT
*              FIELDS, RIGHT- JUSTIFIED, ON ONE LINE.
*          (3) OBSERVED VALUES, NOV OF THEM, IN 10 EIGHT-DIGIT FIELDS
*              PER LINE, AS MANY LINES AS NEEDED, DECIMAL POINTS
*              REQUIRED.
*          (4) ONE LINE PER DESIRED SOLUTION, WITH NU STARTING
*              VALUES OF UNKNOWN PARAMETERS, IN UP TO 20 THREE-
*              DIGIT FIELDS, DECIMAL POINTS REQUIRED.
          COMMON/ONE/OV(200),NOV
          DIMENSION X(20),TITLE(10)
**READ INPUT, AND PRINT BACK OUT.
          READ 1,TITLE
    1     FORMAT(10A8)
          PRINT 13,TITLE
   13     FORMAT(*1* 10A8)
          READ 2,NOV,NU,NSOL,ISTEP
    2     FORMAT(4I3)
          PRINT 14,NOV,NU,NSOL,ISTEP
   14     FORMAT(/* OBSERVED VALS=*I3*, UNKNOWNS=*I3
         1*, SOLUTIONS=*I3 *, ISTEP=*I3)
          READ 3, (OV(I),I=1,NOV)
    3     FORMAT(10F8.3)
          PRINT 15,(OV(I),I=1,NOV)
   15     FORMAT(* OBS VALS: *10F8.3)
**ONE PASS PER SOLUTION: SETS STARTING VALUES, CALLS
*  MINIMIZATION ROUTINE, PRINTS RESULTS.  MAXFN=MAXIMUM EFFORT
*  TO BE EXPENDED BEFORE QUITTING (NO.OF MAJOR CYCLES OF
*  MINIMIZATION ROUTINE).
```

```
          DO 30 K=1,NSOL
          READ 4, (X(I),I=1,NU)
 4        FORMAT(20F3)
          PRINT 5, (X(I),I=1,NU)
 5        FORMAT(/* START VALS= *20F3.1)
          MAXFN=500
          MAX=MAXFN
          CALL MINIM(X,NU,MAXFN,ISTEP,FX,IER)
          NCYC=MAX-MAXFN
          IF(IER.NE.0)PRINT 8
 8        FORMAT(* EFFORT LIMIT EXCEEDED, QUIT AT *)
 30       PRINT 7, NCYC, FX, (X(I),I=1,NU)
 7        FORMAT(* NO. OF CYCLES=*I4/ * MINIMUM FUNCTION VALUE=*
         1 F10.7/ * SOLUTION: *20F6.3)
          STOP
          END


          SUBROUTINE MINIM(X,NU,MAXFN,ISTEP,FX,IER)
**MINIMIZATION ROUTINE--CALLS SUBROUTINE FUNCT REPEATEDLY TO
*  EVALUATE EFFECTS OF TRIAL CHANGES IN PARAMETERS.  MAKES
*  MOST PROMISING CHANGE, AND REPEATS.  WHEN NO LONGER
*  IMPROVING, REDUCES STEP SIZE BY A FACTOR OF 10 AND TRIES
*  AGAIN. WHEN STEP SIZE LESS THAN .001, OR MAXIMUM EFFORT
*  EXCEEDED, QUITS AND RETURNS TO MAIN PROGRAM.
          DIMENSION X(20),XX(20),CHG(20)
          IER=0
          CALL FUNCT(NU,X,FX)
          STEP=ISTEP
          IF(ISTEP.LE.0)STEP=.1
 100      DO 80 I=1,NU
 80       CHG(I)=STEP
**BEGIN MAJOR CYCLE
 50       BEST=0.
          SBEST=0
          IB=0
**LOOK FOR MOST PROMISING CHANGE
          DO 30 J=1,NU
          DO 20 I=1,NU
 20       XX(I)=X(I)
          XX(J)=XX(J)+.001
          CALL FUNCT(NU,XX,FXX)
          DIF=ABS(FXX-FX)
          IF(DIF.LE.BEST) GO TO 30
          BEST=DIF
          IB=J
```

230

```
          SBEST=1
          IF(FXX.GT.FX)SBEST= -1.
   30     CONTINUE
**MAKE MOST PROMISING CHANGE, AND CHECK TO SEE THAT IT IS AN
*  IMPROVEMENT; IF NOT IMPROVING, GO TO DECREASE STEP SIZE.
          DO 40 I=1,NU
   40     XX(I)=X(I)
          XX(IB)=XX(IB)+CHG(IB)*SBEST
          CALL FUNCT(NU,XX,FXXX)
          IF((FX-FXXX).LT..00000001) GO TO 90
          FX=FXXX
          X(IB)=XX(IB)
          MAXFN=MAXFN-1
          IF(MAXFN.GE.0) GO TO 50
          IER=1
          RETURN
**DECREASE STEP SIZE;  IF TOO SMALL, QUIT.
   90     STEP=STEP*.1
          IF(STEP.LT..001)RETURN
          GO TO 100
          END


          SUBROUTINE FUNCT (NU,X,FX)
**CALCULATES FUNCTION FX FOR A GIVEN SET OF TRIAL VALUES.
          COMMON/ONE/OV(200),NOV
          DIMENSION X(20),CV(200)
**PATH EQUATIONS TO CALCULATE IMPLIED VALUES CV(I) FROM TRIAL
*  VALUES OF PARAMETERS X(I) BEGIN HERE.
          CV(1)=X(1)
          CV(2)=X(2)
          CV(3)=X(1)*X(2)
**PATH EQUATIONS END HERE.
**LEAST SQUARES FUNCTION TO BE MINIMIZED. (SUM OF SQUARED
*  DIFFERENCES BETWEEN OBSERVED VALUES OV AND CALCULATED
*  IMPLIED VALUES CV.)
          FX=0.
          DO 20 I=1,NOV
   20     FX=FX+(OV(I)-CV(I))**2
          RETURN
          END
```

Modifying IPSOL for a Generalized Least Squares Solution

An alternative version of the generalized least squares χ^2 criterion for correlations is given by $\mathbf{z}'\mathbf{A}^{-1}\mathbf{z}$, where $\mathbf{z}$ is a vector of differences between Fisher z transforms of the observed and implied correlations, and $\mathbf{A}$ is the variance-covariance matrix among the implied values (see Rao, Morton, Elston & Yee, 1977, or Steiger, 1980). $\mathbf{A}$ is obtained via the formulas for the correlations among correlations originally derived by Pearson and Filon (1898). If the correlations entering into $\mathbf{z}$ are based on independent groups, this computation takes a particularly simple form: $\mathbf{A}$ becomes diagonal, and its elements are N_i- 3, where N_i is the number of cases in the i th group. This is the inverse of the variance of a z transform, which is $1/(N-3)$, and is independent of the size of z, so that changes in the z's during iteration do not affect it. Thus, the calculation amounts to multiplying the squared differences between the observed and implied z for each group by N-3 for that group, and summing over groups.

This procedure, while used for convenience in the example of Chapter 4 (Table 4-17), is not strictly proper there, because some of the groups--in particular, those from the adoption study--are not independent. Many of the same fathers enter, for example, into the father-adopted-child and father-natural-child correlations. There is some evidence to suggest that under such circumstances the model χ^2s will be somewhat underestimated, though the parameter estimates are little affected (Rao et al., 1977; McGue, Wette & Rao, 1984). In the present case, a rather clumsy LISREL approach gave solutions to the models in lines 3 and 6 of Table 4-17 having χ^2s of 8.84 and 2.68, compared to the 8.47 and 2.60 in the table; the parameter estimates were .76 and -.04 in both cases for the first, and .69, .02, -.18, and .13 compared to .68, .02, -.17, and .14 for the second. These results are certainly in good enough agreement for most practical purposes.

The following modifications to IPSOL will allow it to carry out these calculations: Add a z-transform function $\ln((1+r)/(1-r))$, add a vector of N's to the input and to the common statement, and suitably modify the function evaluation in FUNCT. For computational economy, one can transform the observed correlations to z's and the N's to N-3's once at the start when the data are read in to the executive program. Also, it is a good idea to provide a test for out-of-range r's prior to the z transformation in FUNCT; otherwise, if a trial r should happen to equal or exceed 1.0, it will abruptly shut down the calculations.

These can be implemented by making the following changes in the IPSOL program as given:

In both IPSOL and FUNCT, add ,XN(60) to COMMON statement, and add the following line after DIMENSION:

```
Z(R) = .5*ALOG((1.+R)/(1.-R))
```

In IPSOL, add after statement numbered 15:

```
        READ 16,(XN(I),I=1,NOV)
16      FORMAT (20F4)
        PRINT 17,(XN(I),I=1,NOV)
17      FORMAT(*    NS: *20F4)
        DO 25 I=1,NOV
        XN(I)=XN(I)-3.
        IF(XN(I).LT.0)XN(I)=0
25      OV(I)=Z(OV(I))
```

In FUNCT, replace statement numbered 20 with:

```
        IF(CV(I).GE.1.)CV(I)=.9999
        IF(CV(I).LE.-1.)CV(I)=-.9999
20      FX=FX+(Z(CV(I))-OV(I))**2*XN(I)
```

Add the following line of comment to IPSOL instructions for input, preceding the present item (4), which becomes (5):

```
*               (4) NS, NOV OF THEM, INTEGERS, IN 4-DIGIT FIELDS, 20 PER LINE.
```

And change other comments as necessary.

Appendix C: Derivation of Matrix Version of Path Equations

In this appendix we show how the McArdle-McDonald matrix equation described in Chapter 2 can be derived from the structural-equation translation of a path diagram.

Figure C.1 repeats the path diagram used in the example, with the structural equations given to the right of the figure.

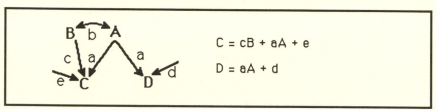

$$C = cB + aA + e$$

$$D = aA + d$$

Fig. C.1 Path model for example.

Now we write out these same equations so that each of them has on its right-hand side all the variables in the path diagram. This means putting in a lot of zero coefficients but gets the equations into a convenient form to express as matrices. For completeness, structural equations have been added for A and B, although these are not very exciting because neither has any incoming causal arrows from other variables in the diagram. In each equation, the first four terms indicate causal arrows from other variables (hence the variable itself always gets a zero coefficient). The last term in each expression is the residual. Source variables such as A and B have by definition all their causes external to the path diagram, so they are treated as "all residual."

$$A = 0A + 0B + 0C + 0D + A$$
$$B = 0A + 0B + 0C + 0D + B$$
$$C = aA + cB + 0C + 0D + e$$
$$D = aA + 0B + 0C + 0D + d$$

Next we write this in matrix form.

$$
\begin{matrix} A \\ B \\ C \\ D \end{matrix}
\quad = \quad
\begin{matrix} 0 & 0 & 0 & 0 \\ 0 & 0 & 0 & 0 \\ a & c & 0 & 0 \\ a & 0 & 0 & 0 \end{matrix}
\quad \times \quad
\begin{matrix} A \\ B \\ C \\ D \end{matrix}
\quad + \quad
\begin{matrix} A \\ B \\ e \\ d \end{matrix}
$$

(You might want to check to assure yourself that the matrix formulation is indeed the equivalent of the equations.)

Now let's call these matrices, from left to right, **v**, **A**, **v** (again), and **u**. The matrix equation can then be written:

$$\mathbf{v} = \mathbf{Av} + \mathbf{u}.$$

By simple matrix algebra, we solve for $\mathbf{v}$:

$$\mathbf{v} - \mathbf{Av} = \mathbf{u}$$
$$(\mathbf{I} - \mathbf{A})\mathbf{v} = \mathbf{u}$$
$$\mathbf{v} = (\mathbf{I} - \mathbf{A})^{-1}\mathbf{u}.$$

If we assume the variables all to be in deviation-score form, the matrix of covariances among all the variables may be obtained as $\mathbf{vv}'/n$, or

$$(\mathbf{I} - \mathbf{A})^{-1}\mathbf{uu}'(\mathbf{I} - \mathbf{A})^{-1\prime}/n.$$

Let us call $\mathbf{uu}'/n$ --a covariance matrix-- $\mathbf{S}$. Then:

$$\mathbf{vv}'/n = (\mathbf{I} - \mathbf{A})^{-1} \mathbf{S} (\mathbf{I} - \mathbf{A})^{-1\prime}.$$

Pre and postmultiplication by $\mathbf{F}$ and $\mathbf{F}'$ selects the observed variables $\mathbf{C}$:

$$\mathbf{C} = \mathbf{F} (\mathbf{I} - \mathbf{A})^{-1} \mathbf{S} (\mathbf{I} - \mathbf{A})^{-1\prime} \mathbf{F}',$$

which is the McArdle-McDonald equation.

Obtaining $(\mathbf{I} - \mathbf{A})^{-1}$

McArdle and McDonald point out that the matrix $(\mathbf{I} - \mathbf{A})^{-1}$ can be obtained for unlooped path diagrams as:

$$\mathbf{I} + \mathbf{A} + \mathbf{AA} + \mathbf{AAA} + \ldots,$$

where the series is carried out until the product terms become zero. If there are no compound paths in the diagram that contain more than one consecutive straight arrow in the same direction, this will occur after the term $\mathbf{A}$; with a maximum of two consecutive straight arrows, it will be after $\mathbf{AA}$; and so on.

The example of Fig. C.1 contains only single straight arrows in its paths. Thus, $\mathbf{AA}$ (as you should verify) is a null matrix. Therefore,

$$(\mathbf{I} - \mathbf{A})^{-1} = \mathbf{I} + \mathbf{A} = \begin{array}{cccc} 1 & 0 & 0 & 0 \\ 0 & 1 & 0 & 0 \\ a & c & 1 & 0 \\ a & 0 & 0 & 1 \end{array}$$

as used for the Chapter 2 example.

In Chapter 1, "direct" and "indirect" causal effects were discussed. Note that matrix $\mathbf{A}$ represents direct effects, and $\mathbf{AA}$, $\mathbf{AAA}$, etc. indirect effects.

Appendix D: Alpha and Canonical Factor Extraction--A Worked Example

In Chapter 5, the Alpha and Canonical rescalings were briefly described. As noted there, Alpha factor analysis rescales variables to equal communalities of 1.0, whereas Canonical factor analysis rescales variables to equal uniquenesses of 1.0.

Formally, the Alpha method rescales variables by $H^{-1}R_rH^{-1}$, where H^{-1} is a diagonal matrix of the reciprocals of the square roots of the commmunalities, and the Canonical method rescales by $U^{-1}R_rU^{-1}$, where U^{-1} contains the reciprocals of the square roots of the uniquenesses. (Recall that one obtains the inverse of a diagonal matrix by taking reciprocals of its elements.) The Alpha rescaling results in a matrix in which differences along the diagonal are eliminated, whereas the Canonical rescaling results in a matrix in which they are enhanced.

Table D-1 illustrates the process with the same reduced correlation matrix used for the centroid and principal factor solutions in Chapter 5; it is shown in the center of the top row. To the left is a column vector of the reciprocals of the square roots of the communalities (e.g., $1/\sqrt{.16} = 2.5$). These are used in the diagonal matrix H^{-1} which pre and postmultiplies R_r to yield the matrix shown on the left in the second row. Note that this has rescaled all the diagonal elements to 1.0, and the other elements proportionately (zeroes, of course, stay zero).

To the right is the Canonical factor solution. At the top are the u^{-1} (e.g., $1-.16 = .84$; $1/\sqrt{.84} = 1.091$). In the second row is the rescaled R_r matrix. Note that the differences are now exaggerated: The high values tend to be scaled up much more than the low values--.16 becomes .19, whereas .91 becomes 10.11. This is because the uniqueness for the first variable is already large (.84) and only needs to be increased a little to equal 1.0, whereas the small uniqueness in the second case (.09) must be increased manyfold.

In the third row of the table are the factor patterns, obtained via VL from the eigenvalue-eigenvector solutions of the matrices above them, as with principal factors. Finally, the factors are returned to their original metric (the standardized variables of the correlation matrix) by the rescalings HP and UP, respectively.

Table D-1 Alpha and Canonical factor solutions (correlation matrix of Table 5-1, exact communalities)

h^{-1}			R$_r$			u^{-1}
2.500	.16	.20	.24	.00	.00	1.091
1.162	.20	.74	.58	.56	.21	1.961
1.348	.24	.58	.55	.41	.21	1.491
1.048	.00	.56	.41	.91	.51	3.333
1.667	.00	.21	.21	.51	.36	1.250

H^{-1}R$_r$H^{-1}					U^{-1}R$_r$U^{-1}				
1.000	.581	.809	.000	.000	.190	.428	.390	.000	.000
.581	1.000	.909	.682	.407	.428	2.846	1.696	3.661	.515
.809	.909	1.000	.580	.472	.390	1.696	1.222	2.037	.391
.000	.682	.580	1.000	.891	.000	3.661	2.037	10.111	2.125
.000	.407	.472	.891	1.000	.000	.515	.391	2.125	.562

Alpha factors			Canonical factors		
I	II	III	I	II	III
.586	.770	.253	.071	.383	.198
.918	.174	-.357	1.344	.992	-.236
.948	.315	.044	.783	.720	.301
.808	-.577	-.115	3.130	-.560	-.038
.694	-.638	.333	.634	-.270	.295

Rescaled Alpha factors			Rescaled Canonical factors		
I	II	III	I	II	III
.23	.31	.10	.06	.35	.18
.79	.15	-.31	.69	.51	-.12
.70	.23	.03	.53	.48	.20
.77	-.55	-.11	.94	-.17	-.01
.42	-.38	.20	.51	-.22	.24

Appendix E: Examples of Input for Factor Extraction by SPSS, BMDP, and SAS

These are examples of inputs to the factor analysis programs of three popular statistical packages. Each will carry out the extraction of three principal factors from the correlation matrix of Table 5-1, using prespecified communalities.

Table E-1 SPSS FACTOR

RUN NAME	CHAPTER 5 EXAMPLE			
VARIABLE LIST	D E F G H			
N OF CASES	100			
FACTOR	VARIABLES=D E F G H/			
	TYPE=PA1/			
	DIAGONAL=.16,.74,.55,.91,.36/			
	NFACTORS=3/			
	ROTATE=NOROTATE			
OPTIONS	3			
STATISTICS	2,3,4,5			
READ MATRIX				
1.00	.20	.24	.00	.00
.20	1.00	.58	.56	.21
.24	.58	1.00	.41	.21
.00	.56	.41	1.00	.51
.00	.21	.21	.51	1.00
FINISH				

Table E-2 BMDP4M

```
/PROBLEM   TITLE='CHAPTER 5 EXAMPLE'.
/INPUT     VAR=5.
           FORMAT='(5F5.2)'.
           TYPE=CORR.
/FACTOR    METH=PFA.
           NUMB=3.
           CONST=.001.
           COMM=.16,.74,.55,.91,.36.
           ITER=1.
/ROTATE    METH=NONE.
/PRINT     INV.RESI.NO SHADE.
           LINE=72.
/END.
  1.00   .20   .24   .00   .00
   .20  1.00   .58   .56   .21
   .24   .58  1.00   .41   .21
   .00   .56   .41  1.00   .51
   .00   .21   .21   .51  1.00
```

Table E-3 SAS PROC FACTOR

```
DATA EXAMP(TYPE=CORR);
_TYPE_ ='CORR';
INPUT _NAME_ $ D E F G H;
CARDS;
D 1.0 . . . .
E .20 1.0 . . .
F .24 .58 1.0 . .
G .00 .56 .41 1.0 .
H .00 .21 .21 .51 1.0
PROC FACTOR METHOD=PRIN NFACT=3;
  PRIORS .16 .74 .55 .91 .36;
  TITLE CHAPTER 5 EXAMPLE;
PROC PRINT;
```

Appendix F: Data Matrix for Thurstone's Box Problem

Table F-1 contains the raw data for the Thurstone box example in Chapter 6.

Table F-1 Data matrix for 40 boxes on 10 variables (after Kaiser & Horst, 1975, all scores x 10)

Box	V1	V2	V3	V4	V5	V6	V7	V8	V9	V10
1	89	50	27	56	34	21	11	8	3	23
2	94	45	38	54	81	34	11	6	4	98
3	69	102	7	103	42	32	11	11	0	99
4	122	96	41	78	61	66	11	11	9	145
5	88	91	92	90	96	82	11	12	11	286
6	182	16	1	96	37	20	14	7	-1	154
7	167	47	30	70	77	36	14	7	7	201
8	169	110	3	118	45	29	14	11	1	60
9	185	87	40	137	77	48	14	10	8	231
10	158	89	104	118	116	94	14	11	11	401
11	137	158	2	149	47	39	14	14	-1	180
12	160	144	41	134	77	85	14	14	8	293
13	164	140	93	143	117	117	15	15	11	492
14	265	36	16	100	43	17	16	7	0	157
15	253	41	37	99	101	26	15	6	7	186
16	229	107	38	143	122	56	17	11	7	351
17	252	74	92	150	138	104	17	11	12	486
18	256	158	4	194	48	43	16	14	0	210
19	249	160	33	226	87	73	17	14	7	432
20	253	165	85	196	151	132	16	13	11	604
21	97	29	18	52	27	28	11	7	0	48
22	91	39	45	57	65	33	11	7	7	124
23	81	80	10	101	28	16	10	12	-1	77
24	83	75	39	83	62	63	12	10	9	148
25	88	108	95	91	96	97	12	12	11	315
26	159	29	7	88	41	16	13	7	1	50
27	166	57	42	86	94	45	14	6	7	223
28	157	90	6	117	47	28	14	10	0	95
29	122	85	35	112	74	63	14	12	7	265
30	171	80	101	114	130	93	14	11	12	393

Table F-1 (continued)

Box	V1	V2	V3	V4	V5	V6	V7	V8	V9	V10
31	174	163	18	173	39	54	15	14	-1	200
32	158	148	49	158	80	88	14	14	7	383
33	165	142	86	146	101	114	14	14	12	445
34	246	46	13	113	52	4	15	7	-1	107
35	263	38	58	87	110	37	16	6	9	207
36	215	85	41	149	101	65	16	10	6	311
37	231	87	83	144	144	82	17	12	13	483
38	235	176	21	185	48	35	16	14	-2	168
39	252	159	43	199	90	78	16	14	7	418
40	248	157	83	199	139	124	16	14	13	554

Note: The boxes are ordered in the table by increasing X dimension; boxes 21-40 represent a repetition of boxes 1-20, with different errors.

Appendix G: Table of Chi Square

df	P= .99	.90	.70	.50	.30	.10	.05	.01
1	.00	.02	.15	.46	1.07	2.71	3.84	6.64
2	.02	.21	.71	1.39	2.41	4.60	5.99	9.21
3	.12	.58	1.42	2.37	3.66	6.25	7.82	11.34
4	.30	1.06	2.20	3.36	4.88	7.78	9.49	13.28
5	.55	1.61	3.00	4.35	6.06	9.24	11.07	15.09
6	.87	2.20	3.83	5.35	7.23	10.64	12.59	16.81
7	1.24	2.83	4.67	6.35	8.38	12.02	14.07	18.48
8	1.65	3.49	5.53	7.34	9.52	13.36	15.51	20.09
9	2.09	4.17	6.39	8.34	10.66	14.68	16.92	21.67
10	2.56	4.86	7.27	9.34	11.78	15.99	18.31	23.21
11	3.05	5.58	8.15	10.34	12.90	17.28	19.68	24.72
12	3.57	6.30	9.03	11.34	14.01	18.55	21.03	26.22
13	4.11	7.04	9.93	12.34	15.12	19.81	22.36	27.69
14	4.66	7.79	10.82	13.34	16.22	21.06	23.68	29.14
15	5.23	8.55	11.72	14.34	17.32	22.31	25.00	30.58
16	5.81	9.31	12.62	15.34	18.42	23.54	26.30	32.00
17	6.41	10.08	13.53	16.34	19.51	24.77	27.59	33.41
18	7.02	10.86	14.44	17.34	20.60	25.99	28.87	34.80
19	7.63	11.65	15.35	18.34	21.69	27.20	30.14	36.19
20	8.26	12.44	16.27	19.34	22.78	28.41	31.41	37.57
21	8.90	13.24	17.18	20.34	23.86	29.62	32.67	38.93
22	9.54	14.04	18.10	21.34	24.94	30.81	33.92	40.29
23	10.20	14.85	19.02	22.34	26.02	32.01	35.17	41.64
24	10.86	15.66	19.94	23.34	27.10	33.20	36.42	42.98
25	11.52	16.47	20.87	24.34	28.17	34.38	37.65	44.31
26	12.20	17.29	21.79	25.34	29.25	35.56	38.88	45.64
27	12.88	18.11	22.72	26.34	30.32	36.74	40.11	46.96
28	13.56	18.94	23.65	27.34	31.39	37.92	41.34	48.28
29	14.26	19.77	24.58	28.34	32.46	39.09	42.56	49.59
30	14.95	20.60	25.51	29.34	33.53	40.26	43.77	50.89
z	-2.33	-1.28	-.52	.00	.52	1.28	1.64	2.33

Note: Table entries are values of χ^2 exceeded P proportion of the time, for given df. For df greater than 30, the quantity $[2\chi^2]^{1/2} - [2df-1]^{1/2}$ may be evaluated as a normal deviate (bottom row). Adapted with permission of Macmillan Publishing Company from *Statistical Methods for Research Workers* (14th Ed.) by R. A. Fisher. Copyright ©1970 by University of Adelaide.

Answers to Exercises

Chapter 1

1. & 2. Various legitimate diagrams are possible, depending on assumptions made--for example, those shown in Fig. H.1.

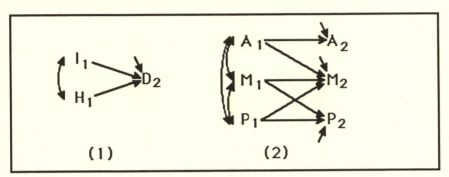

Fig. H.1 Problems 1 & 2--possible answers.

3. Source variables: A, B, W, X, Y, Z. Downstream variables: C, D, E, F, G.

4. That it is completely determined by A and B.

5. $r_{AF} = ae + bf + hcf$; $r_{DG} = cdg + bhdg$; $r_{CE} = ahd$; $r_{EF} = dcf + dhbf + dhae$.

6. $s^2_C = a^2 + i^2$; $s^2_D = b^2 + c^2 + 2bhc$; $s^2_F = e^2 + f^2 + 2eabf + 2eahcf + j^2$.

7. No. There are $(4 \times 3)/2 = 6$ observed correlations, and 8 unknowns (excluding residuals)--a, b, c, d, e, f, g, h.

8. $c_{CD} = a^* s^2_A b^* + a^* c_{AB} c^*$
   $c_{FG} = e^*a^* c_{AB} d^*g^* + f^*b^* c_{AB} d^*g^* + f^*c^* s^2_B d^*g^*$
   $c_{AG} = c_{AB} d^*g^*$
   $s^2_G = g^{*2} s^2_E + l^{*2} s^2_Z$ [or] $g^{*2} k^{*2} s^2_Y + g^{*2} d^{*2} s^2_B + l^{*2} s^2_Z$
   $s^2_D = b^{*2} s^2_A + c^{*2} s^2_B + 2b^*c^* c_{AB}$

9. $D = bA + cB$; $E = dB + kY$; $F = eC + fD + jX$.

10. $r_{BC} = c + ba = .70$; $r_{CD} = a^2 + cba = .48$; $r_{BD} = ba = .30$.
   $a = .6; b = .5; c = .4$ [or] $a = -.6; b = -.5; c = .4$.
   $d = \sqrt{(1 - .36)} = .8$; $e = \sqrt{(1 - .36 - .16 - .24)} = .49$.

11. ab x bc/ac = b^2 = .42 x .14/.12 = .49; b = .7.
     a = .6; c = .2; [or] b = -.7; a = -.6; c = -.2.

12. Centroid factor loadings: .604, .673, .805, .503. Communality estimates--
.36, .44, .63, .25; column sums--1.56, 1.74, 2.08, 1.30; $\Sigma\Sigma$ = 6.68; Σ = 2.5846.

13. [The path model will be: d s_{A2}^2 + bc s_{B1}^2 + ac cov$_{A1B1}$.]

Chapter 2

1. Equations: CV(1)=X(1)*X(2)   Input: PROBLEM 1 CHAPTER 2
              CV(2)=X(1)*X(3)   3 3 2
              CV(3)=X(2)*X(3)   .30   .40   .35
                                          .5 .5 .5
                                          .1 .1 .1
Solutions: a = .585, b = .512, c = .684 (or all negative). Reproduced correlations,
to 3 places, equal to originals (e.g., .585 x .512 = .300). Residual paths: x = .859,
y = .811, z = .729 (e.g., $\sqrt{[1-.512^2]}$ = .859).

2. For example:

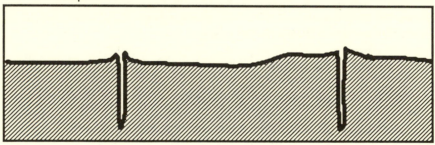

Fig. H.2 An example of a difficult terrain for IPSOL.

3. **A**   **S**

	A	B	C	D
A	0	0	0	0
B	a	0	0	0
C	0	b	0	0
D	0	c	0	0

	A	B	C	D
A	1	0	0	0
B	0	y^2	0	0
C	0	0	x^2	0
D	0	0	0	z^2

F

	A	B	C	D
A	1	0	0	0
C	0	0	1	0
D	0	0	0	1

4. **PH, GA, PS, BE** = 1 x 1; **LX** = 3 x 1; **LY** = 2 x 1; **TD** = 3 x 3; **TE** = 2 x 2.

5.

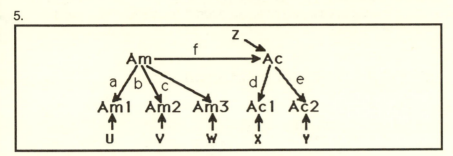

Fig. H.3 Path model for Problem 5.

a = .920, b = .761, c = .652, d = .879, e = .683, f = .356; residual variances: U= .153, V = .420, W = .575, X = .228, Y = .534, Z = .873. χ^2 = 5.80, 4 df, p > .20. The model is consistent with the data. It implies that a little more than one-third (.356) of ambition translates into achievement, when both are expressed in standard-score units.

6. Model 1 (χ^2 = 16.21, 7df) is a significantly poor fit to the data, and a significantly worse fit than any of the other three (χ^2_{diff} 8.09, 2df; 13.71, 3df; 16.13, 6df). None of the others can be rejected (p > .05 for each), but the third fits significantly better than the second (χ^2_{diff} 5.62, 1df).

7.

model	χ^2	unknowns	df	χ^2/df	nfi	pfi	AIC
null	25.00	0	10	2.50			-12.50
1.	16.21	3	7	2.32	.352	.246	-11.10
2.	8.12	5	5	1.62	.675	.338	-9.06
3.	2.50	6	4	.62	.900	.360	-7.25
4.	.08	9	1	.08	.997	.100	-9.04

Model 4 in absolute terms is a very close fit (nfi = .997), but involves many parameters and is probably overfitted, if this is exploratory (χ^2/df<< 1.0). Model 3 is the most parsimonious fit, by either *pfi* or AIC, although one might be a little suspicious of the goodness of the fit (χ^2/df < 1.0) if it were arrived at after much trial and error.

8. LISREL matrices:

	LX	LY	PH	GA	PS	TD	TE	
	1	1	v	a	z	0	x	0
		c					0	y

Unstandardized: $c^* = 1.333$, $v = 1.00$, $a^* = .300$, $z = .173$, $x = .737$, $y = .533$.
Standardized: $b = .512$, $c = .683$, $a = .586$, $z = .657$, $x = .737$, $y = .533$.
(v, z, x, and y are variances, not paths). Standardized solution is in good agreement with IPSOL.

9. $s_A = s_C = s_D = 1.0$

$s_B = \sqrt{(.300^2 \times 1.0 + .173)} = .5128$ [or] $\sqrt{(1.0 - .737)} = .5128$

$a = a^* \times 1.0/.5128 = .300 / .5128 = .585$
$b = b^* \times .5128/1.0 = 1.0 \times .5128 = .513$
$c = c^* \times .5128/1.0 = 1.333 \times .5128 = .684$
path z = path $z^* \times 1.0/.5128 = \sqrt{.173}/.5128 = .811$; variance $= .811^2 = .658$.

Chapter 3

1.

2. First 3 equations, corresponding to r_{12}, r_{13}, and r_{14}:
     CV(1) = X(1)*X(2)
     CV(2) = X(1)*X(3)
     CV(3) = X(1)*X(10)*X(4)
(unknowns are in the same order as in Fig. 3.4).
IPSOL solution: .73, .64, .58, .74, .66, .67, .78, .71, .61, .63, -.41, -.25; in good agreement with Table 3-4 solution (maximum discrepancy $\approx$ .03).

3. One way of setting up LISREL matrices:
LX 1.0   **PH** a   **TD** (diag) b b b [parallel]
        1.0   **TD** (diag) b c d [tau-equivalent]
        1.0
Goodness of fit (maximum likelihood solution):
     parallel: $\chi^2 = 10.35$, 4 df, $p < .05$
     tau-equivalent: $\chi^2 = 5.96$, 2 df, $p > .05$
Reject hypothesis that tests are parallel; hypothesis of tau-equivalence cannot be rejected (but with this small a sample, this does not mean that it fits very well).

4. Within trait across method: .71, .53, .43, .48, .42, .22, .46, .24, .31; median = .43.
     Within method across trait: .37, -.24, -.14, .37, -.15, -.19, .23, -.05, -.12; median absolute value = .19.
     Across method and trait: .35, -.18, -.15, .39, -.27, -.31, .31, -.22, -.10, .17, -.04, -.13, .36, -.15, -.25, .09, -.04, -.11; median absolute value=.175.
Suggests reasonable convergent and discriminant validity of traits, and not a great deal of influence of measurement method.

Answers

5. One way of setting up LISREL matrices (both trait and method factors):

LX
```
a 0 0 j 0 0
0 b 0 k 0 0
0 0 c l 0 0
d 0 0 0 m 0
0 e 0 0 n 0
0 0 f 0 o 0
g 0 0 0 0 p
0 h 0 0 0 q
0 0 i 0 0 r
```

TD (diag): 9 free values

obs. values: $9 \times 10/2 = 45$
unknowns = 30; df = 15

PH
```
1
s 1
t u 1
0 0 0 1
0 0 0 0 1
0 0 0 0 0 1
```

Goodness of fit (maximum likelihood solution):

both kinds of factors	$\chi^2 = 14.15$, 15 df, p > .50
trait only:	$\chi^2 = 21.94$, 24 df, p > .50
method only:	$\chi^2 = 232.77$, 27 df, p < .001

Trait factors, with or without method factors, fit well. Method factors alone do not yield an acceptable fit. Method factors do not add significantly to the fit of trait factors ($\chi^2_{diff} = 7.79$, 9 df, p > .50).

6.

	χ^2	unknowns	df	χ^2/df	nfi	pfi	AIC
null	290.68	9	36	8.07			-154.3
method	232.77	18	27	8.62	.20	.15	-134.4
trait	21.94	21	24	.91	.92	.62	-32.0
both	14.15	30	15	.94	.95	.40	-37.1

The model with both trait and method factors gives the best absolute fit (*nfi* = .95), but the one with trait factors only gives a good absolute fit (*nfi* = .92) and is the most parsimonious by both *pfi* and AIC.

7. One way of setting up LISREL matrices:

LX
```
1  1  1
1  1 -1
1 -1  1
1 -1 -1
```

PH
```
a
b c
d e f
```

TD (diag): g g g g

[second run: same, except c=a]

Trait and acquiescence variances differ: $\chi^2 = 2.61$, 3 df, p > .40;

trait and acquiescence variances the same: $\chi^2 = 156.66$, 4 df, p < .001;

$\chi^2_{\text{diff}} = 154.05$, 1 df, p < .001; reject the hypothesis that trait and acquiescence variances are equal.

8.

CZ	r_{XZ}	r_{XY}	$r_{XY\cdot Z}$
.7	.504	.5184	.3544
.9	.648	.5184	.1698
1.0	.72	.5184	.0
.5	.36	.5184	.4467
.0	.00	.5184	.5184

Only as CZ approaches 1.0 does $r_{XY\cdot Z}$ become small. With values of .5 or less it is only slightly reduced.

9.

Chapter 4

1. Line 2 solution: $\chi^2 = 18.89$, 28 df. Without z for educational aspiration: $\chi^2 = 19.32$, 29 df; $\chi^2_{\text{diff}} = .43$, 1 df, p > .50. Without z for occupational aspiration: $\chi^2 = 30.62$, 29 df; $\chi^2_{\text{diff}} = 11.73$, 1 df, p < .001. Residual correlation of the friends' educational aspirations is not statistically significant, but that for occupational aspirations is.

2. Show that inclusion of curved arrows between the residuals from RPA and REA or ROA, and FPA and FEA or FOA, leads to a significant decrease in χ^2. A problem would be presented by the fact that LISREL does not admit covariances between independent and dependent variables. It would be necessary to reconceptualize the model in order to test it using LISREL.

3. One way of setting up LISREL matrices:

```
LX  a e 0 0 0    PH  1        TD(diag): i j k l i j k l
    b 0 f 0 0
    c 0 0 g 0    χ² = 72.14, 24 df, p < .001;
    d 0 0 0 h    reject such a model.
    a e 0 0 0
    b 0 f 0 0    a = .816, b = .874, c = .172, d = .540,
    c 0 0 g 0    e = .274, f = .259, g = .461, h = .717,
    d 0 0 0 h    i = .259, j = .170, k = .758, l = .194.
```

4. Allow all 8 parameters in the first column of **LX** in #3 to be different. $\chi^2 = 65.53$, 20 df. $\chi^2_{\text{diff}} = 6.61$, 4 df, p > .10. This model does not fit significantly better.

5. With the Jobs path set to 1.0, the χ^2 of 24.56 is the same. The paths corresponding to Table 4-11 are:

'72 B	1.583	.715		'74 B	1.475	.695		'76 B	1.629	.605
	.927	.803			.921	.669			.797	.541
	1.000	1.000			1.000	1.000			1.000	1.000

These values are equivalent to the Table 4-11 values except for scale; that is, for any given year they are proportional--e.g., in 1972 they are in the ratio 1.583/1 for the college group (.927/.58,1/.63); and .715/1 for the noncollege group (.803/1.12, 1/1.40). The standard deviations for the latent variables for the present solution may be obtained as $\sqrt{[1.0(\text{Jobs variance - Jobs residual})]}$; i.e., for 1972, $\sqrt{[1.0(1.67^2-1.744)]} = 1.022$ for the college group, and $\sqrt{[1.0(1.90^2-2.379)]} = 1.110$ for the noncollege group. The standardized path values may then be obtained as 1.583(1.022/2.03) = .80, .927(1.022/1.84) = .51, etc. for the college group, and .715(1.110/1.25) = .63, etc. for the noncollege group, in agreement with the Table 4-11 standardized values.

6. With different structural and measurement models (text), $\chi^2 = 24.56$, with 32 df. Requiring the measurement model to be the same (**LX** and **TD** invariant across groups) yields $\chi^2 = 119.31$ with 56 df; $\chi^2_{\text{diff}} = 94.75$, with 24 df, p < .001. The same measurement model does not fit in both groups.

7. One way of setting up LISREL matrices for the hypothesis of parallel tests is (e.g., for DZ twins):

```
LX   a 0        PH  1              TD  b
     a 0            .5 1               c b
     a 0                               c c b
     0 a                               d e e b
     0 a                               e d e c b
     0 a                               e e d c c b
```

The χ^2 is 91.59 with 79 df, p > .15, and $\chi^2_{\text{diff}} = 16.76$, 10 df, p > .05. Thus, one would not reject the hypothesis that the three scales are parallel tests of numerical ability. (Under this model, the genetic paths *a* are .665, the residual variances *b* are .561, and the residual covariances are *c* = .209 across tests within persons, *d* = .244 for the same test across persons, and *e* = .150 across both tests and persons.)

Chapter 5

1. Eigenvalues: 1.6912,   .9458,   .7866,   .5764

    Eigenvectors: -.601   .109   .282   -.740

                               -.476   .256   -.834   .107

                                 .277   .951   .128   -.037

                               -.579   .133   .456   .663

    Principal Factors: -.781   .106   .250   -.562

                               -.620   .249   -.740   .081

                               .361   .925   .114   -.028

                               -.753   .129   .404   .503

R_1 .61 R_2 .62 R_3 .68 $R_4 = R$

    .48   .38   .51   .45   .33   .99

  -.28   -.22   .13   -.18   .01   .99   -.16   -.08   1.00

    .59   .47   -.27   .57   .60   .50   -.15   .58   .70   .20   -.11   .75

Successive matrices do improve in fit, but much of this improvement, after R_1, is in fitting the diagonal elements.

2. R_r .49   .28   -.14   .42   Eigenvalues: 1.0500, .0000, .0000, .0000

            .28   .16   -.08   .24 v_1 .683 p_1 .700

        -.14   -.08   .04   -.12   .390   .400

          .42   .24   -.12   .36   -.195   -.200

                                      .586   .600

PP' reconstructs R_r exactly.

3. Row sums: 1.26   .88   .48   1.20

    Factor loadings: .645   .450   -.246   .614

4. SMCs: .962, .800, .962, .791, .842

5. Eigenvalues: 2.867, 1.798, .217, .098, .019.   Two factors by either criterion.

6. C_r 11388427

                 62   3

       4140969   335   1481764

        174064   143   72785   10455

        438118   9858   948323   571210   34152979

P from **R**$_r$	.625	.764	**P** from **C**$_r$	.043	.980
	.707	-.550	(rescaled)	.937	-.026
	.711	.679		.152	.980
	.882	-.161		.859	.409
	.741	-.579		.918	-.020

The two solutions are quite different, due to the markedly different scales on which the solution was carried out.

7. First factor loadings: .977, .137, .984, .530, .132. $\chi^2 = 243.74$, 10df, $p < .001$. Reject hypothesis of one common factor.

8.

Chapter 6

1.	**P**	.79	.06	**F**	1.00	.57	**S**	.82	.51
		.89	-.06		.57	1.00		.85	.45
		.07	.68					.45	.72
		-.08	.61					.26	.56

2.

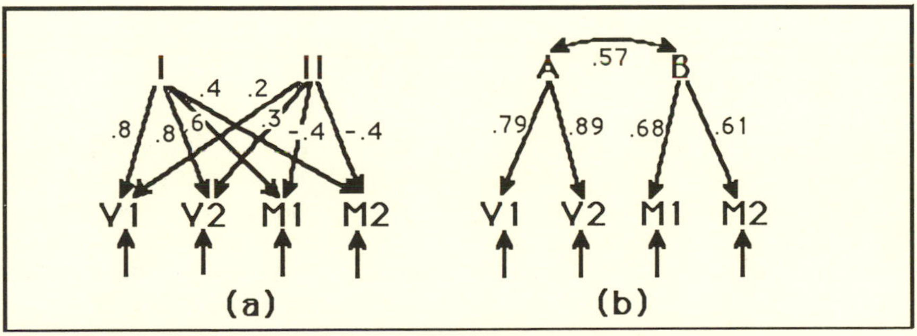

Fig. H.4 Path diagrams for unrotated (a) and rotated (b) factors.

3. Path diagram h^2s: .62, .79, .46, .37.
   h^2s from **P**$_0$ rows: .68, .73, .52, .32.

The two are similar but not identical. The h^2s from the path diagram may be either larger or smaller because some of the small omitted paths are positive and some negative, resulting in covariance contributions of both signs.

4. $P_{varimax}$

.78	.28
.83	.19
.28	.67
.11	.55

$P_{oblimin}$

.79	.07
.88	-.04
.10	.66
-.05	.59

P_{orthob}

.78	.07
.88	-.05
.06	.68
-.09	.61

$F_{varimax}$

1.00	.00
.00	1.00

$F_{oblimin}$

1.00	.52
.52	1.00

F_{orthob}

1.00	.56
.56	1.00

Either oblique **P** is similar to the Problem 1 solution. The orthogonal **P** has similar high loadings, but its low loadings are not as close to zero.

5.

6. P_{promax}

.77	.09
.86	-.02
.07	.68
-.08	.60

F

1.00	.53
.53	1.00

Quite similar to other oblique solutions.

7. P_{-1}

.06	.09	.50
.05	.04	.46
.08	.63	.38
.72	.05	.07
.06	.74	.04
.72	.02	.04

P_0

.01	.01	.58
.01	-.00	.50
.01	.65	.33
.74	.02	.01
.02	.81	-.06
.75	-.02	-.01

$P_{+.5}$

-.28	-.18	.98
-.24	-.26	.95
-.52	1.04	.33
1.32	-.34	-.34
-.40	1.53	-.48
1.37	-.40	-.37

F

1.00	.26	.33
	1.00	.50
		1.00

F

1.00	.41	.49
	1.00	.65
		1.00

F

1.00	.89	.91
	1.00	.94
		1.00

With increasing w, the factors become more highly correlated. Quartimin, $w = 0$, gives the cleanest solution in this case--the small loadings are quite close to zero and split between positive and negative. $w = -1.0$ gives a passable but probably slightly too orthogonal a solution--the low loadings are larger than for the quartimin solution, and all are positive. The solution for $w = +.5$ has clearly gone much too oblique--the factor intercorrelations are approaching unity, the low loadings have become substantial and negative, and the high loadings are implausibly high, yielding an essentially uninterpretable solution.

8. W

.04	.00	.27
.03	-.00	.20
.11	.63	.55
.46	.02	.07
.02	.35	-.05
.43	.00	.05

The scores of the individual would be -.25, +.63, and +.97 on the three factors.

9. Add a row .17 .55 to **S** and a row -.14 .62 to **P**.

10.

Chapter 7

1. A single-factor model with factor pattern the same for both sexes, but latent variable mean and variance and residual variances allowed to differ, fits the data quite adequately: $\chi^2 = 9.25$, 10 df, $p > .50$ (by LISREL with maximum likelihood criterion). Allowing the factor patterns to differ between men and women does not significantly improve the fit: $\chi^2 = 4.61$, 7 df; $\chi^2_{diff} = 4.64$, 3 df, $p \approx .20$. (If in the first condition the residuals are required also to be equal, the fit is still satisfactory: $\chi^2 = 15.72$, 14 df, $p > .30$; but additionally constraining equality of either the latent variable mean or variance leads to significant worsening of fit: $\chi^2 = 456.13$ or $\chi^2 = 24.56$, each with 15 df, χ^2_{diff} $p < .01$.)

2. (a) $C_{S1D2} =$ $aV_{M1}d + eV_{M2}h + iV_{F1}l + mV_{F2}p$
$+ aC_{M1M2}h + eC_{M2M1}d + aC_{M1F1}l + iC_{F1M1}d$
$+ aC_{M1F2}p + mC_{F2M1}d + eC_{M2F1}l + iC_{F1M2}h$
$+ eC_{M2F2}p + mC_{F2M2}h + iC_{F1F2}p + mC_{F2F1}l$

(b) $C_{S1D2} =$ .15 + .10 + .08 + .04 + .15 + .036 + .02 + .024 + .02 + .003
$+$ .004 + .02 + .024 + .015 + .080 + .01 = .776

$C_{S1D1} =$ .779

(c)

A	**B**	**C**	**D**	C_M	C_F	C_{MF}
.5 .2	.5 .3	.4 .1	.3 .2	1.0 .6	1.0 .5	.2 .1
.2 .4	.3 .5	.1 .2	.2 .4	.6 1.0	.5 1.0	.1 .3

$C_{SD} = .779$ .776 C_{S1D1} and C_{S1D2} agree with top
      .579   .651   row of C_{SD}.

3. $V_A = V_X + V_S$ $C_{AC,AD} = hV_{XZ} + hV_{ZS}$
   $C_{B,AC} = 0$ $V_Y = c^2V_X + d^2V_Z + 2cdi + eV_{XZ} + V_W$

4. (a) e.g., $\underline{C \quad D \quad 7 \ S}$, $\Sigma = 15$,   stress = .387
         2.0   2.0 1.0

(b) e.g., $\underline{C \quad D \qquad 7 \ S}$, $\Sigma = 11$,   stress = .332
        2.0   3.0   1.0

5. P   .13 .98
         -.91 -.09
         -.16 -.98
         .82 .04
         .86 .39

Large coefficients on the first factor correspond to equations that give relatively large weights to X, and on the second, to equations that give relatively large weights to X^2.

6. P_{12} .56
          .74
          .88

7. $P_{02} = P_{01}P_{12}$ .52
                  .45
                  .78
                  .44
                  .56
                  .40

8. P

				h^2	h^2_{PF}
.52	.01	.01	.28	.35	.34
.45	.01	-.00	.24	.26	.25
.78	.01	.44	.16	.83	.83
.44	.61	.01	.00	.57	.58
.56	.02	.55	-.03	.62	.61
.40	.62	-.01	-.00	.54	.55

The two sets of communalities agree within rounding error.

9. P   .778 Al   Eigenvalues: 3.12, .67, .15, .06
      -.999 Ben   (factor solution--principal factors
       .943 Carl   with iteration for communalities,
      -.643 Zach starting from SMCs)

The data are fairly well described by a single factor, on which Carl and Al are alike and opposite to Ben and Zach.

10. For example, one might obtain a measure of motor skill for a sample of persons under a number of different conditions, and intercorrelate and factor the conditions to study the major dimensions of their influence on motor skill.

11. It would perhaps better be described as a three-mode analysis carried out in two groups (college and noncollege). The three modes are persons, occasions ('72, '74, '76), and attitudes (toward busing, criminals, jobs).

12.

References

Acock, A. C., & Fuller, T. D. (1985). Standardized solutions using LISREL in multiple populations. *Sociological Methods and Research, 13*, 551-557.

Anderson, J. C., & Gerbing, D. W. (1984). The effect of sampling error on convergence, improper solutions, and goodness-of-fit indices for maximum likelihood confirmatory factor analysis. *Psychometrika, 49*, 155-173.

Baker, R. L., Mednick, B., & Brock, W. (1984). An application of causal modeling techniques to prospective longitudinal data bases. In S. A. Mednick, M. Harway & K. M. Finello (Eds.), *Handbook of longitudinal research. Vol. I: Birth and childhood cohorts* (pp.106-132). New York: Praeger.

Baumrind, D. (1983). Specious causal attributions in the social sciences: The reformulated stepping-stone theory of heroin use as exemplar. *Journal of Personality and Social Psychology, 45*, 1289-1298.

Bentler, P. M. (1980). Multivariate analysis with latent variables: Causal modeling. *Annual Review of Psychology, 31*, 419-456.

Bentler, P.M. (1983a). Some contributions to efficient statistics in structural models. *Psychometrika, 48*, 493-517.

Bentler, P. M. (1983b). Simultaneous equation systems as moment structure models. *Journal of Econometrics, 22*, 13-42.

Bentler, P. M. (1986). Structural modeling and *Psychometrika:* An historical perspective on growth and achievements. *Psychometrika, 51*, 35-51.

Bentler, P. M., & Bonett, D. G. (1980). Significance tests and goodness of fit in the analysis of covariance structures. *Psychological Bulletin, 88*, 588-606.

Bentler, P. M., & Huba, G. J. (1979). Simple minitheories of love. *Journal of Personality and Social Psychology, 37*, 124-130.

Bentler, P. M., & Lee, S.-Y. (1979). A statistical development of three-mode factor analysis. *British Journal of Mathematical and Statistical Psychology, 32*, 87-104.

Bentler, P. M., & Lee, S.-Y. (1983). Covariance structures under polynomial constraints: Applications to correlational and alpha-type structural models. *Journal of Educational Statistics, 8*, 207-222.

Bentler, P. M., & McClain, J. (1976). A multitrait-multimethod analysis of reflection-impulsivity. *Child Development, 47*, 218-226.

Bentler, P. M., & Weeks, D. G. (1980). Linear structural equations with latent variables. *Psychometrika, 45*, 289-308.

Bentler, P. M., & Weeks, D. G. (1985). Some comments on structural equation models. *British Journal of Mathematical and Statistical Psychology, 38*, 120-121.

Bentler, P. M., & Woodward, J. A. (1978). A head start reevaluation: Positive effects are not yet demonstrable. *Evaluation Quarterly, 2*, 493-510.

References

Bock, R. D., Dicken, C., & Van Pelt, J. (1969). Methodological implications of content-acquiescence correlation in the MMPI. *Psychological Bulletin, 71*, 127-139.

Bohrnstedt, G. W., & Felson, R. B. (1983). Explaining the relations among children's actual and perceived performances and self-esteem: A comparison of several causal models. *Journal of Personality and Social Psychology, 45*, 43-56.

Bollen, K. A., & Jöreskog, K. G. (1985). Uniqueness does not imply identification. *Sociological Methods and Research, 14*, 155-163.

Boomsma, A. (1982). The robustness of LISREL against small sample sizes in factor analysis models. In K. G. Jöreskog & H. Wold (Eds.), *Systems under indirect observation* (Part I, pp.149-174). Amsterdam: North-Holland.

Boomsma, A. (1985). Nonconvergence, improper solutions, and starting values in LISREL maximum likelihood estimation. *Psychometrika, 50*, 229-242.

Boomsma, D. I., & Molenaar, P. C. M. (1986). Using LISREL to analyze genetic and environmental covariance structure. *Behavior Genetics, 16*, 237-250.

Bracht, G. H., & Hopkins, K. D. (1972). Stability of educational achievement. In G. H. Bracht, K. D. Hopkins, & J. C. Stanley (Eds.) *Perspectives in educational and psychological measurement* (pp. 254-258). Englewood Cliffs, NJ: Prentice-Hall.

Bramble, W. J., & Wiley, D. E. (1974). Estimating content-acquiescence correlation by covariance structure analysis. *Multivariate Behavioral Research, 9*, 179-190.

Brenner, S.-O., & Bartell, R. (1984). The teacher stress process: A cross-cultural analysis. *Journal of Occupational Behaviour, 5*, 183-195.

Browne, M. W. (1977). Generalized least-squares estimators in the analysis of covariance structures. In D. J. Aigner & A. S. Goldberger (Eds.), *Latent variables in socio-economic models* (pp. 205-266). Amsterdam: North-Holland.

Browne, M. W. (1984). Asymptotically distribution-free methods for the analysis of covariance structures. *British Journal of Mathematical and Statistical Psychology, 37*, 62-83.

Burt C. (1917). *The distributions and relations of educational abilities.* London: London County Council.

Burt, R. S. (1981). A note on interpretational confounding of unobserved variables in structural equation models. In P. V. Marsden (Ed.), *Linear models in social research* (pp. 299-318). Beverly Hills, CA: Sage

Busemeyer, J. R., & Jones, L. E. (1983). Analysis of multiplicative combination rules when the causal variables are measured with error. *Psychological Bulletin, 93*, 549-562.

Campbell, D. T., & Fiske, D. W. (1959). Convergent and discriminant validation by the multitrait-multimethod matrix. *Psychological Bulletin, 56*, 81-105.

References

Carroll, J. D., & Chang, J.-J. (1970). Analysis of individual differences in multidimensional scaling via an N-way generalization of "Eckart-Young" decomposition. *Psychometrika, 35,* 283-319.

Cattell, R. B. (1952). The three basic factor-analytic research designs--their interrelations and derivatives. *Psychological Bulletin, 49,* 499-520.

Cattell, R. B. (1966a). The scree test for the number of factors. *Multivariate Behavioral Research, 1,* 245-276.

Cattell, R. B. (1966b). The data box: Its ordering of total resources in terms of possible relational systems. In R. B. Cattell (Ed.), *Handbook of multivariate experimental psychology* (pp. 67-128). Chicago: Rand McNally.

Cattell, R. B. (1978). *The scientific use of factor analysis.* New York: Plenum Press.

Cattell, R. B., & Cross, K. P. (1952). Comparison of ergic and self-sentiment structures found in dynamic traits by R- and P-techniques. *Journal of Personality, 21,* 250-271.

Cattell, R. B., & Jaspers, J. (1967). A general plasmode (No. 30-10-5-2) for factor analytic exercises and research. *Multivariate Behavioral Research Monographs, 67-3,* 1-211.

Cattell, R. B., & Muerle, J. L. (1960). The "maxplane" program for factor rotation to oblique simple structure. *Educational and Psychological Measurement, 20,* 569-590.

Cattell, R. B., & Sullivan, W. (1962). The scientific nature of factors: A demonstration by cups of coffee. *Behavioral Science, 7,* 184-193.

Chatterjee, S. (1984). Variance estimation in factor analysis: An application of the bootstrap. *British Journal of Mathematical and Statistical Psychology, 37,* 252-262.

Cliff, N. (1983). Some cautions concerning the application of causal modeling methods. *Multivariate Behavioral Research, 18,* 115-126.

Coan, R. W. (1959). A comparison of oblique and orthogonal factor solutions. *Journal of Experimental Education, 27,* 151-166.

Cronbach, L. J. (1984). A research worker's treasure chest. *Multivariate Behavioral Research, 19,* 223-240.

Cudeck, R., & Browne, M. W. (1983). Cross-validation of covariance structures. *Multivariate Behavioral Research, 18,* 147-167

Davidon, W. C. (1975). Optimally conditioned optimization algorithms without line searches. *Mathematical Programming, 9,* 1-30.

Davison, M. L. (1983). *Multidimensional scaling.* New York: Wiley.

DeFries, J. C., Johnson, R. C., Kuse, A. R., McClearn, G. E., Polovina, J., Vandenberg, S. G., & Wilson, J. R. (1979). Familial resemblance for specific cognitive abilities. *Behavior Genetics, 9,* 23-43.

de Leeuw, J., Keller, W. J., & Wansbeek, T. (Eds.). (1983). Interfaces between econometrics and psychometrics. *Journal of Econometrics, 22,* 1-243.

Denison, D. R. (1982). Multidimensional scaling and structural equation modeling: A comparison of multivariate techniques for theory testing. *Multivariate Behavioral Research, 17,* 447-470.

References

Dijkstra, T. (1983). Some comments on maximum likelihood and partial least squares methods. *Journal of Econometrics, 22*, 67-90.

Dixon, W. J. (Ed.). (1983). *BMDP statistical software.* Berkeley, CA: University of California Press.

Donaldson, G. (1983). Confirmatory factor analysis models of information processing stages: An alternative to difference scores. *Psychological Bulletin, 94*, 143-151.

Dong, H.-K. (1985). Non-Gramian and singular matrices in maximum likelihood factor analysis. *Applied Psychological Measurement, 9*, 363-366.

Duncan, O. D. (1966). Path analysis: Sociological examples. *American Journal of Sociology, 72*, 1-16.

Duncan, O. D. (1975). *Introduction to structural equation models.* New York: Academic Press.

Duncan, O. D., Haller, A. O., & Portes, A. (1968). Peer influences on aspirations: A reinterpretation. *American Journal of Sociology, 74*, 119-137.

Dwyer, J. H. (1983). *Statistical models for the social and behavioral sciences.* New York: Oxford University Press.

Endler, N. S., Hunt, J. McV., & Rosenstein, A. J. (1962). An S-R inventory of anxiousness. *Psychological Monographs, 76*, (Whole No. 536).

Etezadi-Amoli, J., & McDonald, R. P. (1983). A second generation nonlinear factor analysis. *Psychometrika, 48*, 315-342.

Fredricks, A. J., & Dossett, D. L. (1983). Attitude-behavior relations: A comparison of the Fishbein-Ajzen and the Bentler-Speckart models. *Journal of Personality and Social Psychology, 45*, 501-512.

Fox, J. (1980). Effect analysis in structural equation models: Extensions and simplified methods of computation. *Sociological Methods and Research, 9*, 3-28.

Fox, J. (1985). Effect analysis in structural equation models II: Calculation of specific indirect effects. *Sociological Methods and Research, 14*, 81-95.

Gerbing, D. W., & Anderson, J. C. (1985). The effects of sampling error and model characteristics on parameter estimation for maximum likelihood confirmatory factor analysis. *Multivariate Behavioral Research, 20*, 255-271.

Goldberger, A. S. (1971). Econometrics and psychometrics. *Psychometrika, 36*, 83-107.

Goldberger, A. S., & Duncan, O. D. (Eds.). (1973). *Structural equation models in the social sciences.* New York: Seminar Press.

Gorsuch, R. L. (1983). *Factor analysis* (2nd ed.). Hillsdale, NJ: Lawrence Erlbaum Associates.

Gottfredson, D. C. (1982). Personality and persistence in education: A longitudinal study. *Journal of Personality and Social Psychology, 43*, 532-545.

References

Guttman, L. (1954). A new approach to factor analysis: The radex. In P. F. Lazarsfeld (Ed.), *Mathematical thinking in the social sciences* (pp. 258-348). Glencoe, IL: Free Press.

Haller, A. O., & Butterworth, C. E. (1960). Peer influences on levels of occupational and educational aspiration. *Social Forces, 38*, 289-295.

Hambleton, R. K., & van der Linden, W. J. (Eds.). (1982). Advances in item response theory and applications. *Applied Psychological Measurement, 6*, 373-492.

Harman, H. H. (1976). *Modern factor analysis* (3rd ed., rev.). Chicago: University of Chicago Press.

Harris, C. W. (1962). Some Rao-Guttman relationships. *Psychometrika, 27*, 247-263.

Harris, C. W., & Kaiser, H. F. (1964). Oblique factor analytic solutions by orthogonal transformations. *Psychometrika, 29*, 347-362.

Heise, D. R. (1975). *Causal analysis.* New York: Wiley.

Hendrickson, A. E., & White, P. O. (1964). Promax: A quick method for rotation to oblique simple structure. *British Journal of Mathematical and Statistical Psychology, 17*, 65-70.

Hinman, S., & Bolton, B. (1979). *Factor analytic studies 1971-1975.* Troy, NY: Whitston Publishing.

Hoelter, J. W. (1983). The analysis of covariance structures: Goodness-of-fit indices. *Sociological Methods and Research, 11*, 325-344.

Huba, G. J., & Bentler, P. M. (1982). On the usefulness of latent variable causal modeling in testing theories of naturally occurring events (including adolescent drug use): A rejoinder to Martin. *Journal of Personality and Social Psychology, 43*, 604-611.

Huba, G. J., & Bentler, P. M. (1983). Test of a drug use causal model using asymptotically distribution free methods. *Journal of Drug Education, 13*, 3-14.

Huba, G. J., & Harlow, L. L. (1983). Comparison of maximum likelihood, generalized least squares, ordinary least squares, and asymptotically distribution free parameter estimates in drug abuse latent variable causal models. *Journal of Drug Education, 13*, 387-404.

Hunter, J. E., Gerbing, D. W., & Boster, F. J. (1982). Machiavellian beliefs and personality: Construct invalidity of the Machiavellianism dimension. *Journal of Personality and Social Psychology, 43*, 1293-1305.

Hurley, J. R., & Cattell, R. B. (1962). The Procrustes program: Producing direct rotation to test a hypothesized factor structure. *Behavioral Science, 7*, 258-262.

IMSL (1984). *IMSL user's manual.* Houston, TX: author.

James, L. R., Mulaik, S. A., & Brett, J. M. (1982). *Causal analysis: Assumptions, models, and data.* Beverly Hills: Sage.

James, L. R., & Singh, B. K. (1978). An introduction to the logic, assumptions, and basic analytic procedures of two-stage least squares. *Psychological Bulletin, 85*, 1104-1122.

Jennrich, R. I., & Sampson, P. F. (1966). Rotation for simple loadings. *Psychometrika, 31*, 313-323.

Jöreskog, K. G., & Lawley, D. N. (1968). New methods in maximum likelihood factor analysis. *British Journal of Mathematical and Statistical Psychology, 21*, 85-96.

Jöreskog, K. G., & Sörbom, D. (1979). *Advances in factor analysis and structural equation models.* Cambridge, MA: Abt Books.

Jöreskog, K. G., & Sörbom, D. (1984) *LISREL VI: User's guide* (3rd ed.). Mooresville, IN: Scientific Software, Inc.

Judd, C. M., & Milburn, M. A. (1980). The structure of attitude systems in the general public: Comparisons of a structural equation model. *American Sociological Review, 45*, 627-643.

Kaiser, H. F. (1958). The varimax criterion for analytic rotation in factor analysis. *Psychometrika, 23*, 187-200.

Kaiser, H. F., & Caffrey, J. (1965). Alpha factor analysis. *Psychometrika, 30*, 1-14.

Kaiser, H. F., & Horst, P. (1975). A score matrix for Thurstone's box problem. *Multivariate Behavioral Research, 10*, 17-25.

Kaiser, H. F., & Madow, W. G. (1974). The KD method for the transformation problem in exploratory factor analysis. Paper presented to the Psychometric Society, Palo Alto, CA, March 28.

Kenny, D. A. (1979). *Correlation and causality.* New York: Wiley.

Kenny, D. A., & Judd, C. M. (1984). Estimating the nonlinear and interactive effects of latent variables. *Psychological Bulletin, 96*, 201-210.

Kiiveri, H., & Speed, T. P. (1982). Structural analysis of multivariate data: A review. In S. Leinhardt (Ed.), *Sociological methodology 1982* (pp. 209-289). San Francisco, Jossey-Bass.

Kim, J.-O., & Ferree, G. D., Jr. (1981). Standardization in causal analysis. *Sociological Methods and Research, 10*, 187-210.

Kroonenberg, P. M. (1983). Annotated bibliography of three-mode factor analysis. *British Journal of Mathematical and Statistical Psychology, 36*, 81-113.

Kroonenberg, P. M., Lammers, C. J., & Stoop, I. (1985). Three-mode principal components analysis of multivariate longitudinal organizational data. *Sociological Methods and Research, 14*, 99-136.

Kruskal, J. B., & Wish, M. (1978). *Multidimensional scaling.* Beverly Hulls, CA: Sage.

Land, K. C., & Felson, M. (1978) Sensitivity analysis of arbitrarily identified simultaneous-equation models. *Sociological Methods and Research, 6*, 283-307.

LaRocco, J. M. (1983). Job attitudes, intentions, and turnover: An analysis of effects using latent variables. *Human Relations, 36*, 813-825.

Lawley, D. N., & Maxwell, A. E. (1971). *Factor analysis as a statistical method* (2nd ed.). London: Butterworths.

References

Lazarsfeld, P. F. (1950). The logical and mathematical foundation of latent structure analysis. In S. A. Stouffer, L. Guttman, E. A. Suchman, P. F. Lazarsfeld, S. A. Star, & J. A. Clausen, *Measurement and prediction* (pp. 362-412). Princeton, NJ: Princeton University Press.

Lazarsfeld, P. F., & Henry, N. W. (1968). *Latent structure analysis.* New York: Houghton Mifflin.

Lee, S.-Y. (1986). Estimation for structural equation models with missing data. *Psychometrika, 51,* 93-99.

Lee, S.-Y., & Jennrich, R. I. (1984). The analysis of structural equation models by means of derivative free nonlinear least squares. *Psychometrika, 49,* 521-528.

Li, C.-C. (1975). *Path analysis: A primer.* Pacific Grove, CA: Boxwood Press.

Lindsay, P., & Knox, W. E. (1984). Continuity and change in work values among young adults: A longitudinal study. *American Journal of Sociology, 89,* 918-931.

Lingoes, J. C. (1973). *The Guttman-Lingoes nonmetric program series.* Ann Arbor, MI: Mathesis Press.

Lingoes, J. C., & Guttman, L. (1967). Nonmetric factor analysis: A rank reducing alternative to linear factor analysis. *Multivariate Behavioral Research, 2,* 485-505.

Loehlin, J. C. (1985). Fitting heredity-environment models jointly to twin and adoption data from the California Psychological Inventory. *Behavior Genetics, 15,* 199-221.

Loehlin, J. C., & Vandenberg, S. G. (1968). Genetic and environmental components in the covariation of cognitive abilities: An additive model. In S. G. Vandenberg (Ed.), *Progress in human behavior genetics* (pp. 261-278). Baltimore: Johns Hopkins Press.

Loehlin, J. C., Willerman, L., & Horn, J. M. (1985). Personality resemblances in adoptive families when the children are late-adolescent or adult. *Journal of Personality and Social Psychology, 48,* 376-392.

Long, J. S. (1983a). *Confirmatory factor analysis: A preface to LISREL.* Beverly Hills, CA: Sage.

Long, J. S. (1983b). *Covariance structure models: An introduction to LISREL.* Beverly Hills, CA: Sage.

MacCallum, R. (1983). A comparison of factor analysis programs in SPSS, BMDP, and SAS. *Psychometrika, 48,* 223-231.

MacCallum, R. (1986). Specification searches in covariance structure modeling. *Psychological Bulletin, 100,* 107-120.

Marsden, P. V. (Ed.). (1981). *Linear models in social research.* Beverly Hills, CA: Sage.

Marsden, P. V., & Campbell, K. E. (1984). Measuring tie strength. *Social Forces, 63,* 482-501.

Marsh, H. W., & Butler, S. (1984). Evaluating reading diagnostic tests: An application of confirmatory factor analysis to multitrait-multimethod data. *Applied Psychological Measurement, 8,* 307-320.

References

Marsh, H. W., & Hocevar, D. (1983). Confirmatory factor analysis of multitrait-multimethod matrices. *Journal of Educational Measurement, 20,* 231-248.

Marsh, H. W., & Hocevar, D. (1985). Application of confirmatory factor analysis to the study of self-concept: First- and higher order factor models and their invariance across groups. *Psychological Bulletin, 97,* 562-582.

Martin, J. A. (1982). Application of structural modeling with latent variables to adolescent drug use: A reply to Huba, Wingard and Bentler. *Journal of Personality and Social Psychology, 43,* 598-603.

Martin, N. G., Jardine, R., & Eaves, L. J. (1984). Is there only one set of genes for different abilities? *Behavior Genetics, 14,* 355-370.

Maruyama, G., & McGarvey, B. (1980). Evaluating causal models: An application of maximum-likelihood analysis of structural equations. *Psychological Bulletin, 87,* 502-512.

Mason, W. M., House, J. S., & Martin, S. S. (1985). On the dimensions of political alienation in America. In N. Tuma (Ed.), *Sociological methodology 1985* (pp.111-151). San Francisco: Jossey-Bass.

Maxwell, A. E. (1977). *Multivariate analysis in behavioural research.* New York: Wiley.

McArdle, J. J. (1980). Causal modeling applied to psychonomic systems simulation. *Behavior Research Methods and Instrumentation, 12,* 193-209.

McArdle, J. J. (1986). Latent variable growth within behavior genetic models. *Behavior Genetics, 16,* 163-200.

McArdle, J. J., & McDonald, R. P. (1984). Some algebraic properties of the Reticular Action Model for moment structures. *British Journal of Mathematical and Statistical Psychology, 37,* 234-251.

McDonald, R. P. (1962). A general approach to nonlinear factor analysis. *Psychometrika, 27,* 397-415.

McDonald, R. P. (1967). Nonlinear factor analysis. *Psychometric Monographs,* No. 15.

McDonald, R. P. (1978). A simple comprehensive model for the analysis of covariance structures. *British Journal of Mathematical and Statistical Psychology, 31,* 59-72.

McDonald, R. P. (1980). A simple comprehensive model for the analysis of covariance structures: Some remarks on applications. *British Journal of Mathematical and Statistical Psychology, 33,* 161-183.

McDonald, R. P., & Mulaik, S. A. (1979). Determinacy of common factors: A nontechnical review. *Psychological Bulletin, 86,* 297-306.

McGue, M., Wette, R., & Rao, D. C. (1984). Evaluation of path analysis through computer simulation: Effect of incorrectly assuming independent distribution of familial correlations. *Genetic Epidemiology, 1,* 255-269.

McIver, J. P., Carmines, E. G., & Zeller, R. A. (1980). Multiple indicators. Appendix in R. A. Zeller & E. G. Carmines, *Measurement in the social sciences* (pp. 162-185). Cambridge: Cambridge University Press.

References

McNemar, Q. (1969). *Psychological statistics* (4th ed.) New York: Wiley.

Mooijaart, A. (1985). A note on computational efficiency in asymptotically distribution-free correlational models. *British Journal of Mathematical and Statistical Psychology, 38,* 112-115.

Morrison, D. F. (1976). *Multivariate statistical methods* (2nd ed.). New York: McGraw-Hill.

Mulaik, S. A. (1972). *The foundations of factor analysis.* New York: McGraw-Hill.

Mulaik, S. A. (1986). Factor analysis and *Psychometrika*: Major developments. *Psychometrika, 51,* 23-33.

Muthén, B. (1983). Latent variable structural equation modeling with categorical data. *Journal of Econometrics, 22,* 43-65.

Muthén, B., & Jöreskog, K. G. (1983). Selectivity problems in quasi-experimental studies. *Evaluation Review, 7,* 139-174.

Neale, M. C., & Fulker, D. W. (1984). A bivariate analysis of fear data on twins and their parents. *Acta Geneticae Medicae et Gemellologiae, 33,* 273-286.

Neuhaus, J. O., & Wrigley, C. (1954). The quartimax method: An analytic approach to orthogonal simple structure. *British Journal of Statistical Psychology, 7,* 81-91.

Newcomb, M. D., & Bentler, P. M. (1983). Dimensions of subjective female orgasmic responsiveness. *Journal of Personality and Social Psychology, 44,* 862-873.

Newton, R. R., Kameoka, V. A., Hoelter, J. W., & Tanaka-Matsumi, J. (1984). Maximum-likelihood estimation of factor structures of anxiety measures: A multiple group comparison. *Educational and Psychological Measurement, 44,* 179-193.

Nie, N. H., Hull, C. H., Jenkins, J. G., Steinbrenner, K., & Bent, D. H. (1975). *SPSS: Statistical package for the social sciences* (2nd ed.). New York: McGraw-Hill.

Overall, J. E., & Klett, C. J. (1972). *Applied multivariate analysis.* New York: McGraw-Hill.

Pearson, K., & Filon, L. N. G. (1898). Mathematical contributions to the theory of evolution. *Philosophical Transactions of the Royal Society (London), Series A, 191,* 229-311.

Rahe, R. H., Hervig, L., & Rosenman, R. H. (1978). Heritability of Type A behavior. *Psychosomatic medicine, 40,* 478-486.

Rao, C. R. (1955). Estimation and tests of significance in factor analysis. *Psychometrika, 20,* 93-111.

Rao, D. C., Morton, N. E., Elston, R. C., & Yee, S. (1977). Causal analysis of academic performance. *Behavior Genetics, 7,* 147-159.

Richardson, M. W. (1938). Multidimensional psychophysics. *Psychological Bulletin, 35,* 659-660. (abstract).

References

Rice, T., Fulker, D. W., & DeFries, J. C. (1986). Multivariate path analysis of specific cognitive abilities in the Colorado Adoption Project. *Behavior Genetics, 16,* 107-125.

Rindskopf, D. (1983). Parameterizing inequality constraints on unique variances in linear structural models. *Psychometrika, 48,* 73-83.

Rindskopf, D. (1984a). Stuctural equation models: Empirical identification, Heywood cases, and related problems. *Sociological Methods and Research, 13,* 109-119.

Rindskopf, D. (1984b). Using phantom and imaginary latent variables to parameterize constraints in linear structural models. *Psychometrika, 49,* 37-47.

Rindskopf, D., & Everson, H. (1984). A comparison of models for detecting discrimination: An example from medical school admissions. *Applied Psychological Measurement, 8,* 89-106.

Rozeboom, W. W. (1978). Estimation of cross-validated multiple correlation: A clarification. *Psychological Bulletin, 85,* 1348-1351.

Rummel, R. J. (1970). *Applied factor analysis.* Evanston, IL: Northwestern University Press.

Ryan, B. F., Joiner, B. L., & Ryan, T. A., Jr. (1985). *Minitab handbook* (2nd ed.). Boston: Duxbury Press.

Saris, W. E., de Pijper, M., & Mulder, J. (1978). Optimal procedures for estimating factor scores. *Sociological Methods and Research, 7,* 85-106.

SAS Institute (1982). *SAS user's guide: Statistics.* Cary, NC: author.

Satorra, A., & Saris, W. E. (1985). Power of the likelihood ratio test in covariance structure analysis. *Psychometrika, 50,* 83-90.

Schmid, J., & Leiman, J. M. (1957). The development of hierarchical factor solutions. *Psychometrika, 22,* 53-61.

Schmitt, N. (1982). The use of analysis of covariance structures to assess beta and gamma change. *Multivariate Behavioral Research, 17,* 343-358.

Schmitt, N., & Stults, D. M. (1986). Methodology review: Analysis of multitrait-multimethod matrices. *Applied Psychological Measurement, 10,* 1-22.

Schoenberg, R., & Richtand, C. (1984). Application of the EM method: A study of maximum likelihood estimation of multiple indicator and factor analysis models. *Sociological Methods and Research, 13,* 127-150.

Sobel, M. E., & Bohrnstedt, G. W. (1985). Use of null models in evaluating the fit of covariance structure models. In N. Tuma (Ed.), *Sociological methodology 1985* (pp. 152-178). San Francisco: Jossey-Bass.

Sörbom, D. (1974). A general method for studying differences in factor means and factor structure between groups. *British Journal of Mathematical and Statistical Psychology, 27,* 229-239.

Spearman, C. (1904). "General intelligence," objectively determined and measured. *American Journal of Psychology, 15,* 201-292.

SPSS (1984). *USERPROC LISREL: Using LISREL within SPSS-X.* Chicago: author.

References

Steiger, J. H. (1980). Tests for comparing elements of a correlation matrix. *Psychological Bulletin, 87*, 245-251.

Stelzl, I. (1986). Changing a causal hypothesis without changing the fit: Some rules for generating equivalent path models. *Multivariate Behavioral Research, 21*, 309-331.

Tanaka, J. S., & Huba, G. J. (1985). A fit index for covariance structure models under arbitrary GLS estimation. *British Journal of Mathematical and Statistical Psychology, 38*, 197-201.

ten Berge, J. M. F., & Knol, D. L. (1985). Scale construction on the basis of components analysis: A comparison of three strategies. *Multivariate Behavioral Research, 20*, 45-55.

Tesser, A., & Paulhus, D. L. (1976). Toward a causal model of love. *Journal of Personality and Social Psychology, 34*, 1095-1105.

Thurstone, L. L. (1947). *Multiple-factor analysis.* Chicago, University of Chicago Press.

Torgerson, W. S. (1958) *Theory and methods of scaling.* New York: Wiley.

Tucker, L. R. (1964). The extension of factor analysis to three-dimensional matrices. In N. Frederiksen and H. Gulliksen (Eds.) *Contributions to mathematical psychology* (pp. 109-127). New York: Holt, Rinehart and Winston.

Tukey, J. W. (1954). Causation, regression and path analysis. In O. Kempthorne, T. A. Bancroft, J. W. Gowen & J. L. Lush (Eds.), *Statistics and mathematics in biology* (pp. 35-66). Ames, IA: Iowa State College Press.

Vandenberg, S. G. (1962). The Hereditary Abilities Study: Hereditary components in a psychological test battery. *American Journal of Human Genetics, 14*, 220-237.

Vogler, G. P. (1985a). *Multivariate path analysis of cognitive abilities in reading-disabled and control nuclear families and twins.* Unpublished PhD dissertation, University of Colorado, Boulder, CO.

Vogler, G. P. (1985b). Multivariate path analysis of familial resemblance. *Genetic Epidemiology, 2*, 35-53.

Vogler, G. P., & DeFries, J. C. (1985). Bivariate path analysis of familial resemblance for reading ability and symbol processing speed. *Behavior Genetics, 15*, 111-121.

Wellhofer, E. S. (1984). To "educate their volition to dance in their chains": Enfranchisement and realignment in Britain, 1885-1950. *Comparative Political Studies, 17*, 3-33, 351-372.

Werts, C. E., & Linn, R. L. (1970). Path analysis: Psychological examples. *Psychological Bulletin, 74*, 193-212.

Werts, C. E., Linn, R. L., & Jöreskog, K. G. (1977). A simplex model for analyzing academic growth. *Educational and Psychological Measurement, 37*, 745-756.

Widaman, K. F. (1985). Hierarchically nested covariance structure models for multitrait-multimethod data. *Applied Psychological Measurement, 9*, 1-26.

References

Wold, H. (1982). Soft modeling: The basic design and some extensions. In K. G. Jöreskog & H. Wold (Eds.), *Systems under indirect observation* (Part II, pp. 1-54). Amsterdam: North-Holland.

Wright, S. (1920). The relative importance of heredity and environment in determining the piebald pattern of guinea-pigs. *Proceedings of the National Academy of Sciences, 6*, 320-332.

Wright, S. (1960). Path coefficients and path regressions: Alternative or complementary concepts? *Biometrics, 16*, 189-202.

Index

Note: Only names are indexed in the main text, subjects also in the Notes.